To My Wife

The author wishes to thank Deborah Burns for her guidance and patience in the preparation of this book.

Cover and text design by Cindy McFarland
Production assistance by Wanda Harper
Front cover photograph by Elinor Mettler
Back cover art by Elayne Sears
Typesetting by Accura Type & Design

TITLE PAGE PHOTOGRAPH COURTESY OF THE INTERNATIONAL ARABIAN HORSE ASSOCIATION
PART ONE TITLE PAGE PHOTOGRAPH COURTESY OF THE AMERICAN QUARTER HORSE ASSOCIATION

Printed in the United States by The Alpine Press

First Printing, April 1989

Library of Congress Cataloging in Publication Data

Mettler, John, J., 1923–
Horse sense : a complete guide to horse selection and care / John J. Mettler. — 1st ed.

"A Garden Way Publishing book."
Includes index.
ISBN 0-88266-549-9
ISBN 0-88266-545-6 (pbk.)
1. Horses. 2. Horses—Selection. I. Title.
SF285.3.M48 1989
636.1—dc19 88-82753

CONTENTS

PREFACE

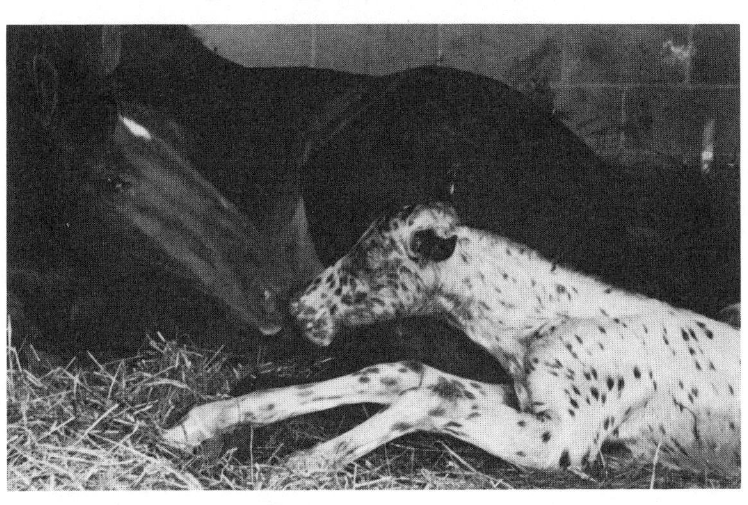

SANDRA FAYE PRUETT/APPALOOSA HORSE CLUB

***As the mare** licked her newborn foal, the gangly youngster struggled to get his long, ungainly legs moving in the right direction. Finally the colt managed to get on his feet, and as he searched out his mother she gently guided him in the right direction until the soft mouth reached a teat. Seconds later he was nursing joyfully.*

"I thought this was her first foal," a bystander remarked. "How does she know how to help him?"

"Instinct," the owner replied. "Mother instinct."

Noticing a young boy petting the nose of one of the other horses in the stable, the bystander said, "I suppose that's instinct, too."

"It sure is! It's an instinct almost as strong as mother love, for humans to love horses."

A tough old trainer, the owner knew people as well as he knew horses. This love of horses seems to be as strong an instinct as any I know of. There are cat lovers and people who hate cats. There are even people who never in their lives felt any affection toward "man's best friend," the dog. But although I know people who are afraid of horses, I have yet to hear anyone say, "I hate horses."

Horses are big business in the United States today. In New York State, for example, the horse industry is second only to the dairy industry in agricultural economic impact. The people who make their living from horses are professionals, capable of knowing where to seek help and advice when they need it. When they call a veterinarian it's for assistance from a purely veterinary medical standpoint.

But then there is the person who, because of the aforementioned instinct, has bought a horse, and knows nothing more about the animal except that he loves it. This kind of new owner not only doesn't know whom to ask for help, but he or she doesn't even know what questions to ask. As a client of mine admitted, "My education in horse care seems to consist of jumping in head first and finding out what has to be done as I naively fumble along."

This book is written for enthusiastic but inexperienced horse owners such as these, as well as for experienced owners who wish to understand their horses better.

PROLOGUE

METTLER

*"**Sorry to bother you** on a Sunday morning, but..."*

I didn't need to ask who it was. Paul Weaver, who owned a riding stable, was my oldest client and a good friend.

"...I've got an extra horse this morning."

"I didn't know you had a mare due to foal this late in the year, Paul," I teased.

"No, honest, Doc, this is no joke. When I went down to feed, there was this beautiful grey roan Arabian filly in my yard. I can't figure out where she came from and thought you might know."

"Any halter on her?"

"Yes," Paul answered, "a brand new halter and about 15 feet of rope."

I thought a minute, then answered, "I can't think of anyone who might stake out a youngster, but if anyone calls I'll send them around."

"She's too valuable a piece of horseflesh not to be missed," Paul added. "In fact, she's too valuable and too young to be tied out. I can't figure it out, but I'll phone some of the other neighbors."

About two hours later the phone rang again.

"Doctor, this is Jim Deane on the old Keating place. Paul Weaver told me to phone you."

I'd heard of Jim, the new young attorney who had joined an old law practice in town. For a moment I wondered if Paul had gone to a lawyer about this lost filly.

Then he continued, "I bought a horse yesterday and during the night she broke loose and wandered down to Paul Weaver's. He told me to get you over right away to check her out. He's afraid she may have overeaten before she got to his place. I found our grain barrel tipped over and at least half a bag of grain gone."

When I arrived at the Deane place I was pleased to see the filly tied to a hitching post eating coarse dry hay. Jim said Paul had told him to do that. As I examined the filly Jim told me his story. He'd grown up in a town down county and had always wanted a horse. But while working his way through high school, college, and law school he could never afford one. His wife, Sue, an accountant by profession, was a city girl and she, too, had always wanted a horse. When they moved to the country they decided it was time to fulfill their lifelong dream.

The filly had belonged to a woman I knew who happened to owe Jim's law firm a large sum of money. The older attorney took the filly, rather than lose everything, and gave it to Jim and Sue. They didn't have a place ready for a horse yet, so they tied her to a stake on the lawn for the night. This morning when they got up she was gone.

The filly was a beauty, with flaxen mane, dappled grey coat, and classic Arabian head. Examination showed all vital signs normal, no indication that the grain she'd eaten was bothering her, and no cuts on her fetlock from the rope, as is usually the case when a green horse is staked out. Still, when I considered the amount of grain that was missing I decided to give her a gallon of light mineral oil. Once the grain started to ferment in her digestive tract it would have been too late.

"Did you get any health papers with her?" I asked Jim.

"No, but the owner said she'd had all her shots in May, and shouldn't need anything until next May."

"What about a Coggins?"

"What's that?"

"A test for Equine Infectious Anemia. Although the disease is now nearly wiped out it's still required on all horses at sale time."

"Oh yes, the woman said it was done in May, but she'd lost the papers."

"That's no problem then," I said. "I know her veterinarian, and I'll ask him for a copy. But just to be sure, in a month we'll retest her. Paul Weaver lost a whole group of horses with E.I.A. twenty years ago and he won't be happy unless we retest this one that will be so close to his stable. I'll call that veterinarian and check on those shots. In the meantime, give her all the hay she wants but no more grain until Tuesday. Restrict her water to half a pail every four hours until you see the oil coming through, which will be sometime tomorrow."

The next morning I phoned the other veterinarian, Dr. Schmidt, and, much to my disappointment, I learned that he hadn't even seen the filly or any other horse on that farm in the last six months. When I phoned Jim at his office to tell him the bad news he said, "Well, go ahead and do the

Coggins. I'd like you to check her out anyway to see if she is worth the money we allowed on her."

"Okay, and I'll give her a tetanus booster at the same time."

"No," Jim said, "I believe the former owner, that she had the shots, and I don't want to put any more money into her than I have to."

That afternoon I met Jim and his wife at their place right after lunch. I took blood for the Coggins and again offered to do a tetanus booster, since it costs very little. Again Jim said no. I checked the filly and she seemed as healthy as could be, although she did need her feet trimmed.

"Are you going to send her to someone to be trained?" I asked.

"No, Sue and I figure we'll learn while she learns. The woman who had her before said she's green broken."

"Where are you going to keep her? I see you haven't completed her yard yet."

"Mrs. Cole up there by the pond is going to let us use her yard and barn. That way we'll have room for a pony for our daughter and eventually another horse for me."

"Good, Jim," I said. "I apologize for seeming nosy, but if I came to you with a legal problem I'd expect you to steer me along with the right questions and advice."

"Thanks," Jim answered, "but we'll be okay. Sue's uncle used to work on a farm when he was a boy. He's retired and lives near here now, so he'll help us."

Time goes fast in the fall, and from October, when the Deanes first got the filly, whom they named Nadia, until December, I didn't hear from them. Paul Weaver told me one day, however, that the filly had run away with Sue, causing her to fall off and injure her back.

"Why, oh why, do people who are beginners buy a horse that is a beginner, too?" Paul exclaimed.

"I don't know, Paul, but maybe it's because if they go into any other kind of livestock, whether it's chickens, rabbits, cattle, or pets such as cats or dogs, they start with young ones."

Christmas Eve was wet and muddy that year, and at five o'clock I returned from calls, tired, with wet feet and cold hands. As I walked in the telephone rang.

"Doc, it's Jim Deane. I just went up to feed Nadia, and Mrs. Cole pointed out to me that she is lame. I picked up her foot and found a staple in it. Mrs. Cole suggested I call you for that tetanus booster."

Christmas Day was a Saturday, so it would be three days if I waited until Monday to give the shot. There wasn't much choice — either give it tonight or it might be too late.

Even in the poor light, when I got the mud off the foot and cleaned it up I could see that the staple had been in the foot several days. The frog, the soft triangular part of the sole, was already loose, as though it had abscessed and broken out. Jim confirmed this. Mrs. Cole told him the filly had been lame several days, but he'd been busy, and was feeding her before daylight in the morning and after dark at night. I trimmed and packed the foot, put on a water-resistant bandage and gave Nadia a tetanus toxoid injection. I said I would stop Monday to remove the bandage and check the foot.

The Deanes went away for Christmas and Mrs. Cole's grandson was to feed and care for Nadia and the pony. Sunday evening Mrs. Cole called me and said her grandson had told her Nadia was having trouble eating and was acting strangely. When I arrived the filly was standing in the stall slobbering in her grain, her tail was half raised and she didn't look like the beautiful graceful creature I was used to seeing when I drove past. As I touched her to put a halter on, she flew back and her "third eyelid," the pink covering of the inner corner of the eye, came up and nearly covered the eye. In all my years of practice I'd seen only five full-blown cases of tetanus, or lockjaw, but this was a textbook case.

We treated Nadia for a week, using antibiotics, muscle relaxants, painkillers and antiserums, feeding her with a stomach tube. Some days she seemed almost normal, but within twelve hours she would be worse again. Jim and Sue had returned and by New Year's Day had all they could stand. Jim had read all he could on the disease and had even talked to a college classmate who was now a professor of veterinary neurology at the university. He learned that tetanus is one of the easiest diseases to prevent, but almost impossible to cure.

"Put her down," Jim said sorrowfully. "We can't stand to see her suffer any more."

The next day Jim telephoned and asked if he and Sue could come to my office for a visit. When they arrived he said, "Remember last October when you told me that if you came to me with a legal problem you'd expect me to steer you with the right questions and answers? Well, I really didn't know what questions to ask you about horses. Sue and I still want a horse. Can you just give me some advice as to what to look for this time?"

"Jim," I said, "you've already made enough mistakes to discourage most people, but they weren't your fault. No one can learn, even in a lifetime, all there is to know about horses. I grew up when horsepower was still used on the farm and after thirty-five years of veterinary practice there is never a day that I don't learn something more. I'd be glad to share that information with you."

Before you buy a horse, or accept one for whom its owner is seeking "a good home," ask yourself if you really want to take on the responsibility and expense of caring for a horse properly. Just loving your pet horse is not enough. Like a child, it needs not only your love, but your constant tender loving care and discipline around the clock.

Before even going out to look for a horse, either take some lessons at a reliable riding academy or spend some time helping out at a stable. Pony club members are taught to muck out a stall, groom a horse, and clean and care for tack before they even start to ride.

The initial purchase price of a horse is only a small part of the cost. From then on there is the everyday cost of grain, hay, bedding, shoeing, and veterinary care, plus the work of feeding, grooming, mucking stalls, and caring for tack. You may find if you take lessons that you really would be just as happy riding once a week on someone else's horse as owning one yourself, or you might even discover that you don't like horseback riding as much as you had expected.

More likely, you will find that you won't be happy until you have a horse of your own. Then, as you gain experience with it, you will probably begin to look forward to the challenge of your next horse. Welcome to the wonderful world of horse ownership.

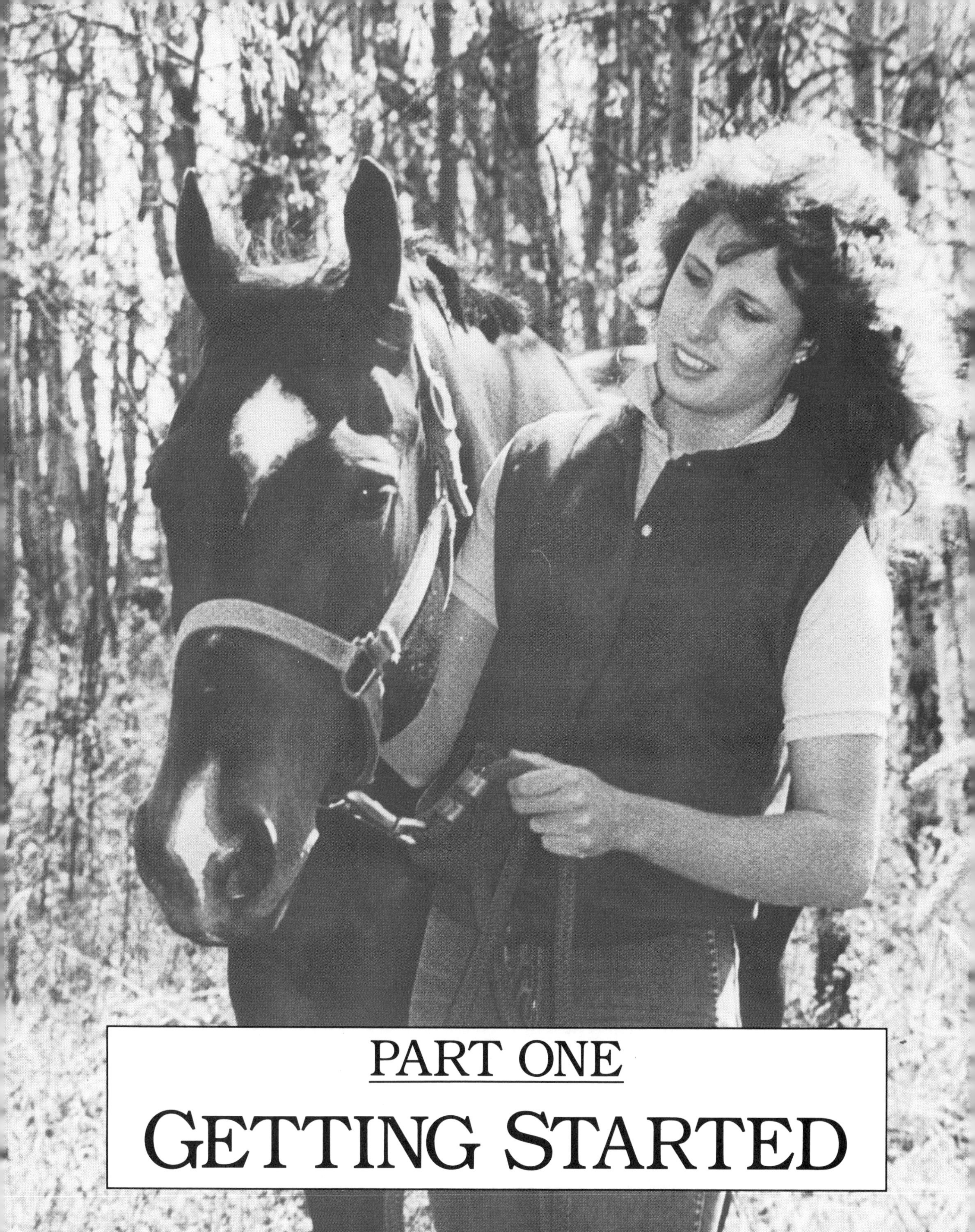

PART ONE
GETTING STARTED

CHAPTER I

INTERNATIONAL ARABIAN HORSE ASSOCIATION

SELECTING YOUR FIRST HORSE

__Much of what I'm telling you,__ others learned by harsh experience. It was hard not only on them and their horse, but also on me, as their veterinarian. The "all I'm looking for is a good home" horse can be a nightmare, as two young clients of mine found out. They had moved into a small town after living many years in the suburbs where they couldn't keep a horse. Their new home had an old barn that had housed horses fifty years ago, and they had a backyard big enough to provide a little exercise yard and access to pasture. Despite what sounds like an ideal set-up, they got into all sorts of difficulty, thanks to an over-solicitous uncle.

Uncle Charlie sent the girls a horse from a riding stable before they'd had time to clean the junk out of the old barn or put a fence around the yard. They moved into their country home one day and the horse arrived the next. Uncle Charlie said the horse, an old Palomino, had belonged to some friends who no longer had use for it and were tired of paying board to the riding stable. "All they wanted was a good home for old Pal," he told them during a telephone call just about an hour before the horse arrived in a broken-down trailer.

Within twelve hours the girls called me. A neighbor who knew horses had noticed that Pal was hung up in what was left of the old barnyard fence and advised them to call. Pal had pushed against the fence, knocking it down, and then, as he tried to back up, had caught his front foot over the wire and cut himself.

It was already dark when I arrived, but even in the beam of a flashlight the old horse appeared more like the one in the Frederic Remington painting "The End of the Trail" than like the product of a good riding stable. I stitched up his leg the best I could, gave him tetanus toxoid, and told the girls to keep Pal in the barn overnight until they fixed the fence in the morning. I said I'd be back in a couple of days to check on him.

The next day they called again. They had straightened up the fence before they turned Pal out, but he was caught again, this time in an old garden cultivator that was hidden in some weeds in back of the barn. Also, Pal wouldn't eat a thing.

Again we patched him up the best we could, but now that it was daylight I was able to look him over a little better. He had a parrot mouth; that is, his lower jaw was shorter than the upper, and what few teeth he had left in back were so out of line that every time he tried to bite his teeth hit bare gums. To further complicate things, seeing him in daylight I realized he had cataracts in both eyes and was totally blind.

When I tried to cut the longer teeth back so they wouldn't injure his gums, making it possible for him at least to eat soft food, he struck at me and I knew I'd have to give him a sedative. Before giving anything in the vein I always check an animal's heart. When I did, I realized that poor old Pal not only was blind, accounting for his walking into things in a strange yard, and not only were his teeth in bad shape, but he had a very bad heart as well. And there was something more — something strange about this old horse that I couldn't put a finger on. I pleaded with the girls, "The most humane thing to do with this poor old fellow is to put him down."

This brought a flood of tears and the accusation that I was cruel and heartless.

"Uncle Charlie promised the owner we'd give Pal a good home," the new owners insisted.

Cautioning them that the sedative might be dangerous, and praying that it wouldn't be, I gave the old horse Rompun and cut off the sharp teeth. When I returned a few days later to remove the stitches from the first wound I was pleased to see that the poor old horse was eating and had apparently learned his way around the fence, because he hadn't cut himself anymore. I took a fecal sample, telling the girls that if it was positive for worms it would be safe to worm the old horse now that he was eating better.

The sample was positive and two days later I stopped to do the worming. The old horse was standing in the sun with his head against the barn, seeming more asleep than awake. Knowing that blind horses are easily frightened, I spoke to him so as to warn him I was near. Pal didn't seem to hear me, so I spoke louder, a good sharp "ho!", and put my hand on his rump. As I touched him he suddenly exploded with both hind feet, missing me by inches.

Then I had the missing clue. We not only had a blind horse with a bad heart and poor teeth, but he was also deaf. I left the girls a note, again warning them of the danger they were in if they kept the horse. He could not be ridden safely and sooner or later he was going to hurt someone badly by falling on them or kicking them.

That evening I received a phone call from the girls' mother asking if I would call Uncle Charlie. Call him I did, and I told him in no uncertain terms of the danger in which he had placed his nieces. Two days later old Pal was painlessly put to sleep.

If you have given all of the foregoing a good deal of careful thought and are still convinced that you want to own a horse, the first question is how to go about selecting the right horse, one especially suited to the first-time horse owner.

Age should be your first consideration. Horses live to be well into their twenties, or even older. At eleven years they are considered aged, yet from then on most have at least ten good, serviceable, trouble-free years ahead before they become too old. Until a horse is seven or eight it is still in the learning stage of its life, able to pick up bad habits as well as good, and still dependent upon the rider to guide it. For this reason, regardless of the age of the owner, a first horse should be a quiet, mature animal, well trained for its intended use. Some horses that were never properly trained, or were trained for a particular purpose, such as parade horses, are always too much for an inexperienced person to handle. Too old and crippled, of course, can be just as bad as too young and flighty.

There is nothing worse than an old horse or pony that has had its own way all of its life. These spoiled animals seem to

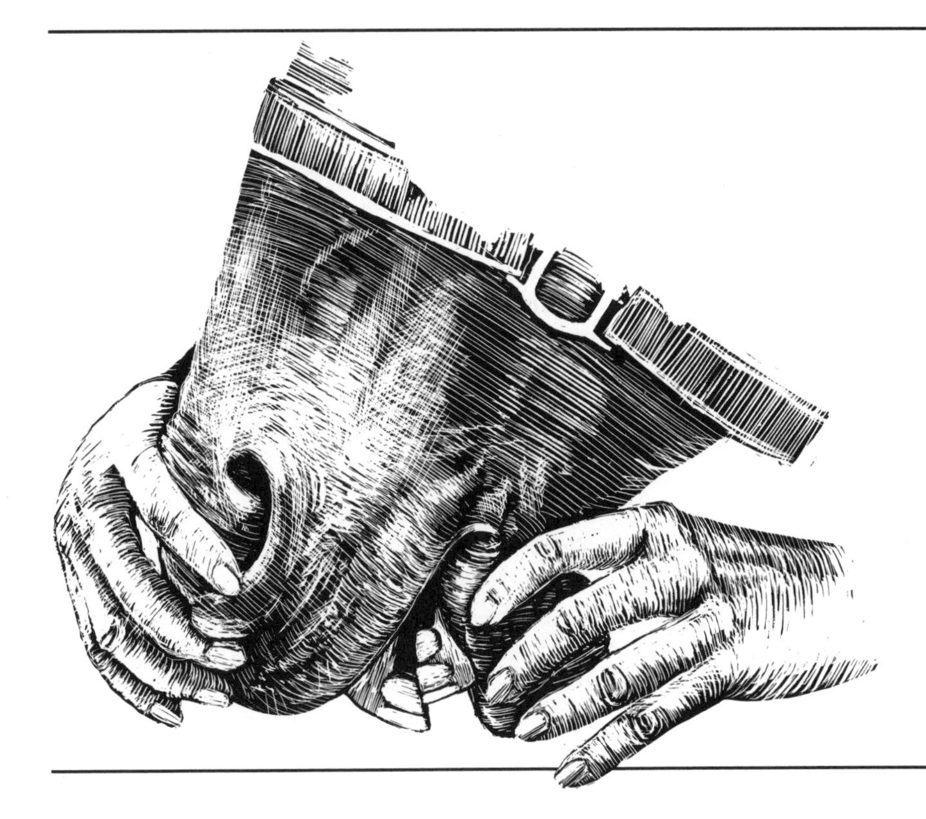

Parrot mouth

know they can get away with anything with an inexperienced person. Raising the head so high that a small person can't bit them, running away or back to the barn, knocking the rider off under low tree limbs, striking and rearing when someone is working around their heads, or not allowing themselves to be caught are the most common problems.

Ponies are not young horses, but breeds of small horses. Technically, any horse under 14.2 hands is a pony (a hand is four inches, so 14.1 hands would be 57 inches at the highest point of the shoulder, or withers), but for our purposes, I shall describe as ponies those bred as ponies, such as the Shetland, Welsh, Pony of America, etc.

Within some horse breeds, such as Arabian, individual animals may never grow big enough to be technically any more than a pony, and some ponies, such as the Connemara, are more horse than pony, both in size and disposition.

For some reason, perhaps because ponies are smarter or more cunning than full-size horses, or because people don't take them seriously enough to train them properly, the bad habits I mentioned are more apt to be noticed in ponies. All too often a small pony is bought for a child, who either becomes discouraged with its downright meanness or outgrows the pony by the time he is old enough and sufficiently experienced to handle it. In general avoid small ponies when choosing the first mount for a small child.

For the older child, however, a really good, well-broken, well-behaved pony is worth its weight in silver. A pony is fine for a family with several children if the older, more experienced members of the family can help the younger ones to ride.

Never buy a stallion (an unaltered male pony or horse) for a first horse. Some stallions, Arabians in particular, are as quiet and well behaved as any gelding when you see them at a show or on the trail, but you never know when their male instinct will suddenly take over. The odor given off by a mare in season or even the odor of blood or perspiration on a human may sometimes excite a stallion and start him biting, striking, or rearing.

A prospective horse owner should learn the correct terminology in reference to age and sex of horses. A newborn horse is a foal, a colt if male and a filly if female. At four years of age the colt is known as a stallion, unless altered or castrated, in which case it is called a gelding. Once four years old, the filly is called a mare.

Before looking for a horse you should decide what you really want. For example, do you want a riding horse (Western or English), a driving horse, one you can ride *and* drive, an all-purpose horse that you can ride, drive, and use as a work horse on a small farm—or do you just want a pet?

Regardless of what you are going to use your first horse for, its disposition is more important than its conformation or appearance. A mature horse, eight years old or over, is more apt to be quiet and easy to get along with. The only way you can find out if you and the horse get along with each other is to try it, and work with it.

Conformation, or how a horse is put together, is important in two ways. First, a horse must have good conformation to live and work a full life without constant lameness and gait problems. Second, and just as important to your satisfaction in ownership, a horse should be pleasing to the eye. Its head should have an attractive appearance. It may sound silly, but your own intuition about a horse, whether it appears kind and intelligent, wild-eyed and flighty, dull and uncooperative, or just plain mean and cranky, is very apt to be correct, even if you are inexperienced in horsemanship.

The illustration on page 6 shows that horses, regardless of breed, should have

CONFORMATION OF A HORSE

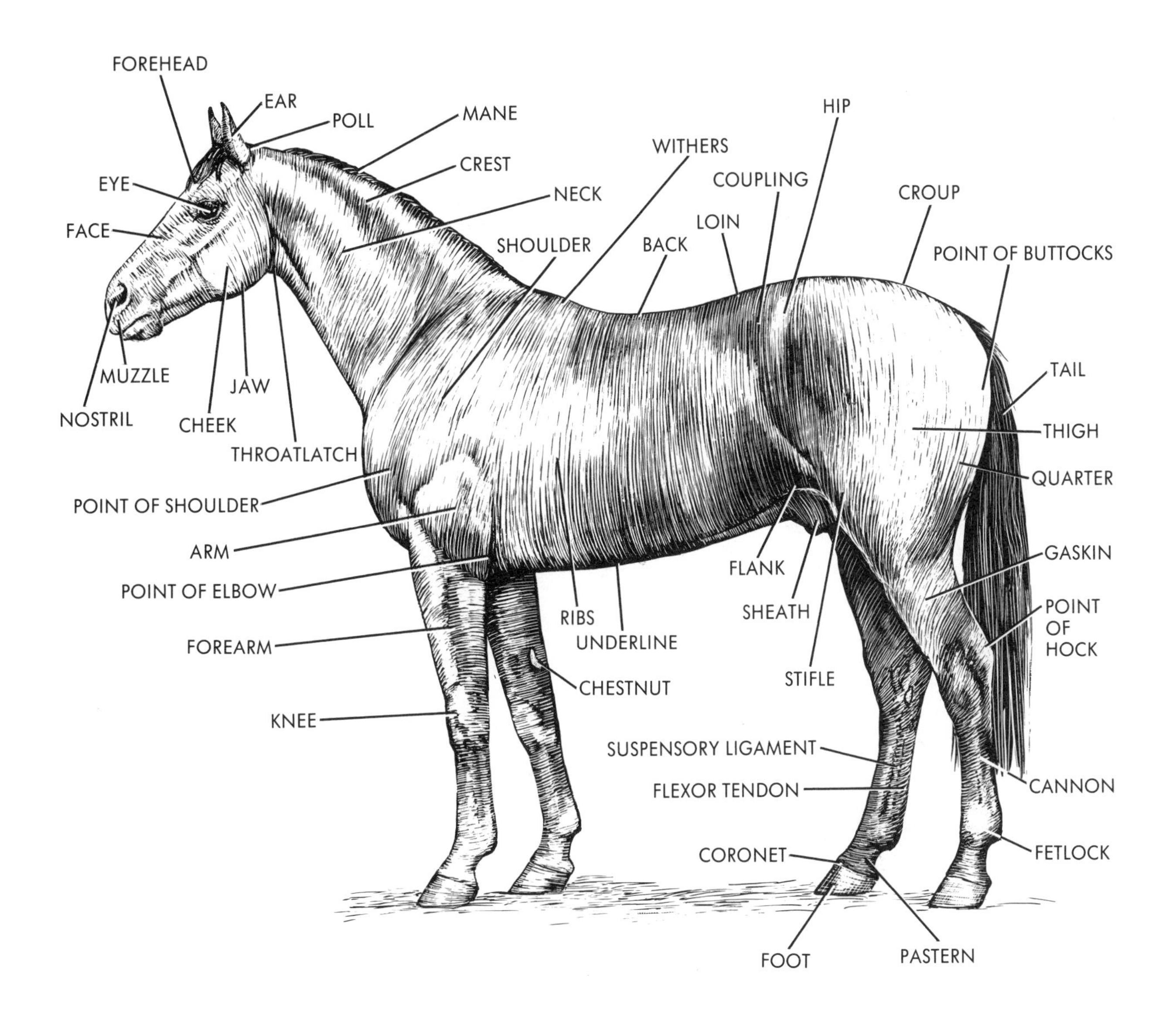

strong, well-placed legs, not crooked but not too straight either. Legs set too close together lead to injuries caused by hooves striking the opposite legs. When legs are too close together at the front of the body, the chest is narrow with too small a space for heart and lungs. Legs too crooked cause joint problems due to straining of tendons and ligaments. Legs too straight can lead to concussion of the joints caused by the jarring of one bone against the other in too straight a line.

Again, as in judging a horse by its head, pleasing appearance of body and legs can tell even a beginner much about the strength and worth of a horse.

Even though a horse's disposition is more important to a first-time horse buyer, there is little joy in owning a horse with a Roman nose, ewe neck (thin curved neck), sway back, goose rump (concave instead of convex), and cow hocks (crooked rear legs with too much angle at the hock).

Don't turn a horse down simply because it is thin and has a rough hair coat, if it has a good disposition. Fifty or 100 pounds of weight in the right places and a shine to the hair coat can improve the appearance of any horse.

Almost as important as the horse itself is the person from whom you buy it. Buy your first horse from, or through, a professional horseman with a good reputation. Buying from a friend, or from someone who offers a horse at a very low price "to give it a good home," too often results in acquiring a useless horse, as well as ending a good friendship.

Once you have found a likely candidate, how do you know if this is "the one"? First try riding or driving the horse at the seller's place. Then, assuming you have a suitable place to keep a horse, make arrangements to take it home on trial for a week or a month. This will usually involve a down payment and certain agreements on both sides, such as that it is your responsibility if the horse gets sick or is injured while in your possession, and that if you don't keep it you will forfeit the down payment and

LEG CONFORMATION

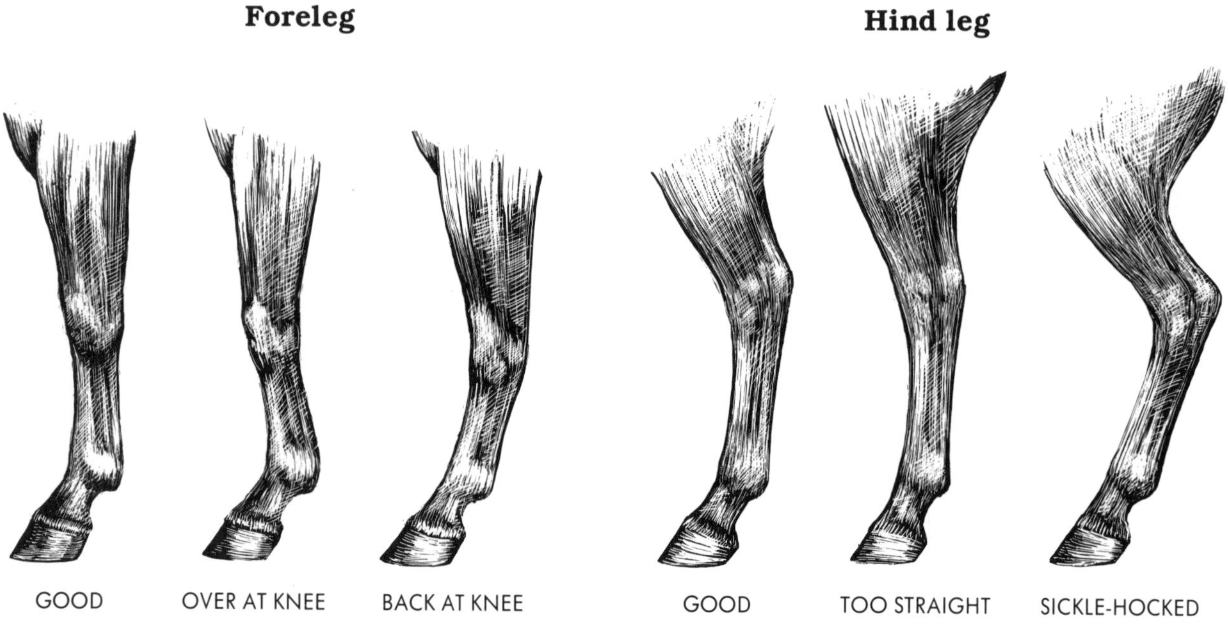

trucking costs. You may be able to buy a horse for less money without a trial, but it's more of a gamble.

Before the horse is taken home on trial, have it checked by an experienced horseman or a veterinarian recommended by a professional horseman. X-rays, blood chemistry, and examination of the respiratory system with an endoscope are not necessary on an ordinary, reasonably priced animal. What a professional can tell about feet, legs, teeth, and eyes, however, will indicate whether or not you have a serviceable horse. A horse that has sight or hearing problems or poor coordination, can be dangerous to the rider and even to itself.

If the seller won't permit you to take the horse on trial or give a thirty-day guarantee, and you are not 100 percent certain that the horse has not been on "bute" or other pain-killers to conceal lameness, or tranquilizers to hide bad habits, ask for a blood test for such things. At the price of horses today you can't be any more trusting when buying a horse than you are when you buy a used automobile.

It would be fine if the seller could furnish you with a complete health history of the horse, but this is almost impossible except on a sale of a young horse by a breeding farm. However, the seller should furnish you with a recent valid negative report on a Coggins test for Equine Infectious Anemia. In some states this is required by law.

Make sure that the horse has had a tetanus booster within the past year.

CHECKLIST FOR SELECTING THE RIGHT HORSE

- **If it is your first horse, it should be a quiet, mature animal, well trained for its intended use.**
- **In general, avoid small ponies when choosing the first mount for a small child.**
- **Never buy a stallion for a first horse.**
- **Buy your first horse from, or through, a professional horseman with a good reputation.**
- **Have your horse checked by a veterinarian before you buy it.**
- **Make sure the horse has had a recent tetanus booster and a Coggins test.**

CHAPTER II

KATHY KADASH/HORSEMAN MAGAZINE

HOUSING THE BACKYARD HORSE

***Veterinarians** get to know every farm in their area, and although sometimes the owner changes, they know where to go when a new owner calls saying he is on "the old so-and-so place on such and such a road." Calls to a village are different, although it is usually easy to see which residences have room for a backyard horse.*

I once spent nearly an hour trying to locate a first-time client in a small town. Although I found the correct street, I drove up and down becoming more and more confused since none of the homes gave any indication that they could possibly have a horse. Knocking on the door of a house with no fences, no outbuilding in its yard, and only a one-car garage, asking, "Are you Mrs. Jones who called about her horse?" can cause a veterinarian to be checked out by the local police as a suspicious character!

After numerous inquiries I finally found a woman who said a new family had just moved in on the next block and rumor had it that they had a horse in the garage.

The reason for the call was to check a horse for a head injury. When I finally found Mrs. Jones she told me that they had just moved to this town from a place 100 miles away. They had brought with them their daughter's Saddlebred mare which, prior to the move, had been kept at a riding stable.

The mare had an injury on her head all right — in fact, she had a swelling the size of a softball right between her ears. Leading her out of the two-car garage was no problem; she seemed aware of the low-cross beams and kept her head at normal level. Bad as the swelling looked, it seemed as though it would get better by itself if it weren't bumped again.

"She only bumps it when she is fed," Mrs. Jones explained. "In fact, we discovered this morning that if she is fed in a different corner she eats quietly."

To prove her point she put grain in a plastic pail and led the mare back into one corner of the garage. There the mare ate normally, but when feed was put in a metal manger in the opposite corner the mare wouldn't even go near it. Coaxed to do so, she began to eat, then suddenly jumped back, striking her poor sore head on a cross beam.

You don't need a veterinarian, Mrs. Jones, I thought, you need an electrician.

I went over and carefully put my hand on the metal manger, but got no shock. Then I wet the back of my hand and touched it to the rim of the manger...still no shock. I was glad I hadn't mentioned electricity, because perhaps there was something else frightening the mare, like a mouse or a bird's nest. I stood pondering, with one hand on the manger. Suddenly I felt a definite electric shock—not enough to hurt, but it was there. I put more pressure on the manger and it moved slightly, giving me a more severe jolt.

The electrician told me later that one of the lag screws holding the manger in place had punctured a metal conduit that came up the wall from an underground electrical service. Pressure on the manger as the horse pushed her nose down made better contact and she jumped back. If she ate very carefully, not pushing down, she felt nothing.

The term "backyard horse" is in common use today, but the average back yard needs some alterations before it really is a suitable place to keep a horse. Besides protection from the weather, a horse needs a place to exercise, and there has to be room to store hay, grain, bedding, and tack.

Before starting a building project, check your local zoning regulations. Even if your neighbor has a horse in his back yard, he may be operating under a "grandfather" clause which permits continuation of a non-conforming use that was in existence prior to passage of the zoning ordinance. Up until fifty years ago most backyards in small-town America had a place for a horse, and despite zoning regulations enacted in recent years a zoning appeals board may grant a variance if public opinion is favorable toward horses. If you follow proper procedures, and have your neighbor's blessing, you can avoid much unpleasantness later.

When starting from scratch, a pole-barn type of shed, with one side open to the south or east, is all that is needed. This type of building can be constructed in a small yard, about a quarter of an acre, where the horse can exercise freely and yet find protection from cold rain, flies, and hot sun when it so desires. Horses stay healthier under these conditions than when kept in a tightly closed barn.

If you are considering converting a building, be sure it is high enough on the inside so that a horse won't injure itself when it throws up its head. This rules out most

Pole-barn shed and horse yard

one- or two-car garages. Concrete floors should also be avoided. A stall for a horse should be at least 12 feet by 12 feet, although 10 feet by 12 feet is often all that can be worked out. Straight stalls, 10 feet by 5 feet, with the horse tied, used to be common for work horses, but are rarely used today for light horses. If you think you may someday want to have a mare foal, one stall in your barn should be at least 14 feet by 14 feet, or 14 feet by 16 feet. Stalls that are too large, however, can lead to a horse developing bad habits when you try to halter it.

Your local Cooperative Extension Service and most feed companies can furnish you with plans for everything from the open shed to a far more elaborate stall-type barn with storage area. Whether you design the building yourself or use professional plans, be sure the grain room is secure. Not only should there be a horse-proof door and latch that closes automatically, but all grain storage containers should also have lids that horses can't open should storm or accident cause the feed room door to come open. Hay can be stored under tarps, but is better stored off the ground and protected from weather by some sort of roof. Overhead hay mows have some advantages, but are a fire hazard, and you have the problem of lifting all the hay up into the mow for storage. Mow areas should be made bird-proof if possible. Bird droppings and feathers will ruin hay and may be a source of disease.

THE HORSE YARD

The horse yard should be enclosed by a board fence. This can be very expensive for a large yard or pasture. For a pasture or yard two acres or more you might use wood fence only in the areas near the barn and where horses are fed and watered, and use wire fence elsewhere. Gates should always be wood or metal. Never use "Texas"-style barbed-wire gates. Once they become fence-wise, horses will not get hurt on a good tight barbed wire fence any quicker than on smooth or page wire which they might put a foot through and get caught. Rubber-coated nylon fencing was thought to be the answer to wire cuts a few years ago, but then horses began to chew on it and die from wads of nylon in their digestive system. Some horses, in fact, will chew on wood fences, but this usually does more harm to the fence than to the horse. However, lead-based paint used on any wood surface a horse can lick or chew is deadly poison.

Horses should never be staked out, as you might have seen cattle and goats tied. The usual result is a rope burn under the fetlock, which is difficult to heal and sometimes leads to permanent lameness.

Electric fencing is fine for cattle, but seldom is practical with horses except for temporary use. It may be useful at the top of a stallion enclosure or as a temporary gate, although I don't like it around horses at all. Use wood where you can; if you have to use wire make it barbed, and keep it tight. For more on fencing, see Chapter IV.

If you are lucky enough to have a stream, pond, or spring at the end of a pasture your only problem with water will be restricting it when horses are hot after use. Freeze-proof waterers are available and should be installed when possible, or have the water piped from your barn to a frost-proof hydrant. In this way you can water from buckets without having to carry them too far.

If you can afford it, try to arrange a place under cover where you can cross-tie a horse while you work on it. Cross-tying means simply having two leads, one from each side of a span at least six feet across, which are attached on opposite sides of the halter. This creates a safe convenient place to do everything from grooming and tacking to having the horse shod or veterinary work done. If a horse has never been in cross-ties it takes a bit of training to get it used to it, but the effort is well worthwhile.

If you are in a small town with close neighbors you may need a covered place to store manure. Horse manure doesn't attract flies or give off as strong an odor as the manure of other animals, but it should be picked up and piled where it will compost. In most areas home gardeners will gladly haul away all the manure you will give them.

CHECKING FOR HAZARDS

Before your horse arrives, and every time you enter the area where it is kept, look for possible hazards. The most common are hooks of hardware or wire, and projections from broken hardware or fencing. Bolts used in building fences or walls should be cut off flush with the surface. Broken or unused gate hinges should be removed. Not only can horses catch a halter on these things and hang themselves, but they can also tear skin. One of the most common injuries to horses, even in well-managed stables, is a torn eyelid. Either from scratching itself or purely by accident, the horse can catch an eyelid over a projecting wire, nail, or water pail hanger, pull back in fright, and rip out a whole section of skin.

A horse can hang itself and die because it catches its halter over a tiny projection of stall hardware. The same thing can happen if it finds a hole, from a loose board, for example, large enough to put its head through, but too small to pull it back quickly. A horse reaching for grass through a fence and around a post can catch a halter on a stub of wood or untrimmed hardware and injure itself, or even hang itself as it pulls back in fright.

Another common hazard in pastures and yards is the old post hole which has not been filled in. Just bulldozing over the holes with a few inches of dirt creates a deadly trap into which a horse can step and fall, breaking a leg. Just as dangerous are wooden posts cut off at ground level. After a few years they decompose, leaving a hole which a horse can fall into. Wooden or metal posts broken off just above the ground are equally dangerous, causing grave injuries to feet and lower legs. Woodchuck burrows in a pasture are also hazardous to horses. The woodchuck must be removed and the holes filled in.

Avoid traps in fences and buildings, such as projecting feed mangers, narrow passageways, and sharp corners. These are particularly dangerous when new horses are turned into a pasture or yard, or if one horse in a group is a bully and will push other horses into such places.

Sometimes such hazards as I've described are not noticed by the person who sees them everyday, but stand out like a red warning signal to a stranger to the area. For this reason, when you have visitors who are experienced horse people, ask them to point out anything like this that they notice. Anyone who has ever had a horse killed or injured will be particularly sensitive to these hazards and will probably point out some I've missed.

HALTERS

One of the most controversial questions in horse care is whether to leave halters on or off. On large farms, and with certain individual animals, halters are left on to facilitate catching them. Also, on many farms halters are left on stabled horses so they can more easily be caught and removed from the stable in case of fire.

There are halters with catches that break if a horse becomes caught in a "trap" such as I've described. Cheap leather halters usually will break away, too, but in either case, if the horse twists the halter doesn't always break. Some people leave nylon halters on their horses, replacing one of the brass fittings with cotton shoelaces which they hope will break in an emergency. For the person with only a few horses I recommend leaving halters off unless a horse is extremely difficult to catch.

CHECKLIST FOR HOUSING THE BACKYARD HORSE

- Check your local zoning regulations.
- If you are converting an existing building into a stable, be sure the ceiling is high enough so that the horse can throw its head up without injuring itself.
- A stall for a horse should be 12 feet by 12 feet, if possible. A mare's foaling stall should be 14 feet by 16 feet.
- Never stake out a horse.
- Be sure the grain room is secure.
- Store hay off the ground and under a roof if possible.
- Make sure the horse yard is enclosed by a board fence.
- Do not use lead-based paint on barn or fencing.
- Avoid barbed-wire gates.
- Provide limited access to frost-proof water.
- Arrange a place to cross-tie the horse under cover.
- Have a place to store manure.
- Check to make sure that stall and yard are free of potential hazards.

CHAPTER III

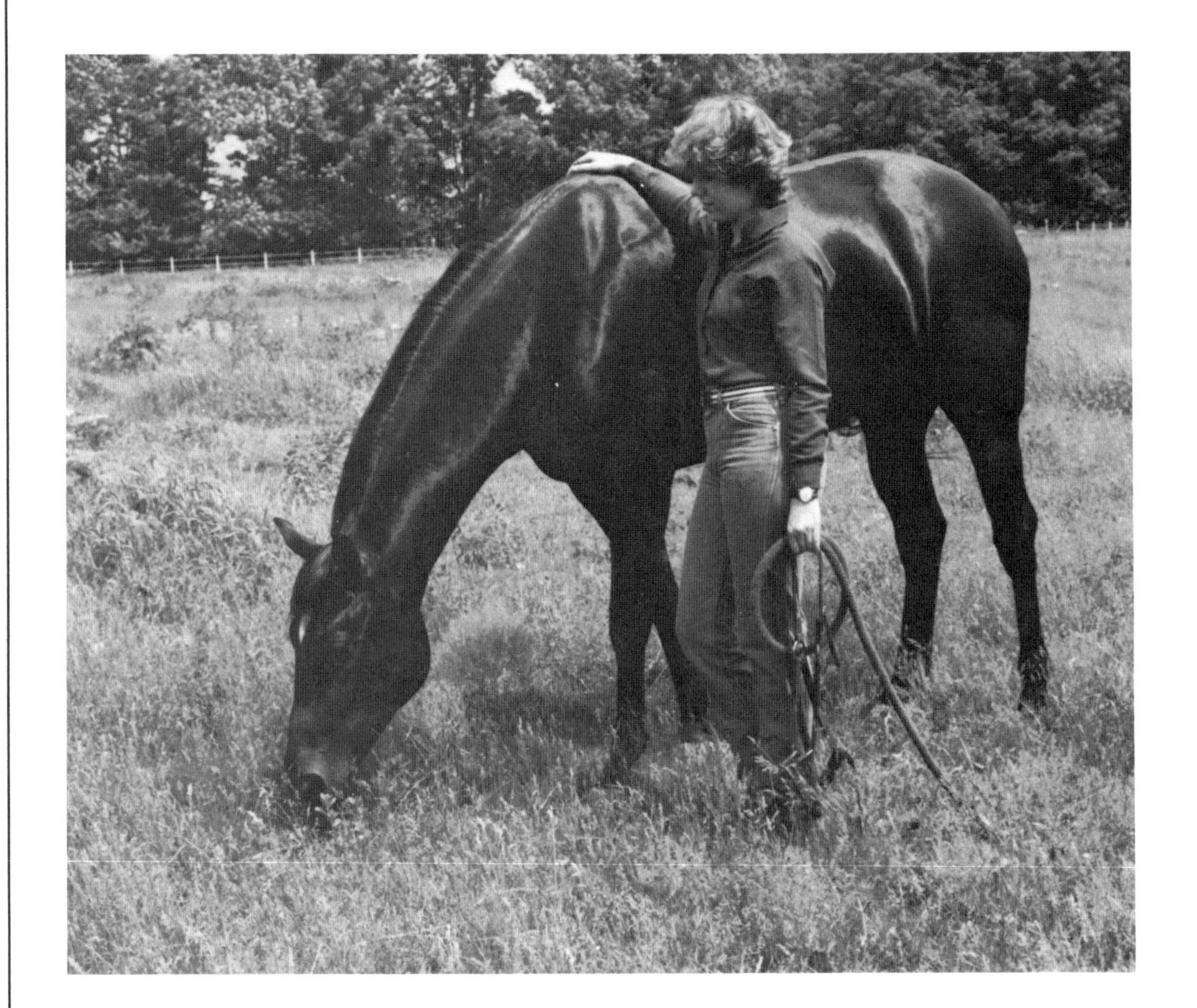

AMERICAN QUARTER HORSE ASSOCIATION

FEEDING AND PASTURE

Charlie Brown was a good pony — everything people wish a pony would be. He loved children, didn't kick, bite, run away, or object to being saddled. He would go slowly for the smallest child, but move out at a brisk trot when asked to by older, more experienced riders. His good manners made him an exception among ponies.

His history was even more unique. A Standardbred trainer saw him one fall day at a horse auction, from which he was bound to end up at "the killer's." She bought him for $25 to save his life and found him so safe and sensible that she gave him to her youngest daughter. When she outgrew him he went to another family. This was repeated three more times over a ten-year period until he ended up on a large dairy farm.

As sometimes happens, the children in this family tired of Charlie, and he wasn't ridden or even fed or petted. The father of the family felt sorry for Charlie and, not being at all experienced with horses, turned him out with the heifers. This was fine for a while — Charlie seemed to like their company and found plenty to eat in the pasture.

Fall came, and grass became scarce. The heifers began to get thin, but Charlie was his usual roly-poly self. As is the case on good dairy farms, hay and grain were brought to the heifers to supplement the short, dry grass. Later, when pasture was completely gone, silage was added to the heifers' diet and, although he had never tasted it before, Charlie found it palatable. Out of boredom, Charlie used to chase the heifers. He found it great fun to charge them like a cow pony, to see them kick up their heels, throw up their tails and run. After they ran, Charlie would stand at the manger and gorge himself on silage and grain.

The following spring the woman who originally bought Charlie turned up at the farm looking for him. She now had a four-year-old grandson and wanted to give him a pony for his birthday. As no one, since she bought him, had ever paid anything for Charlie, she felt she still had a claim on him. If he wasn't being ridden, she told the dairy farmer, she'd like to have him back.

When they went to the heifer yard to look at Charlie he was, as usual, standing at the manger eating. He was in a foot of mud, and was so fat and dirty that he looked like a hairy football with legs. The woman called to him and he looked back at her as though to say, "Am I dreaming?" She called again and he turned to come toward her, slopping through the mud like a broken mechanical toy. As he reached dry ground the woman gasped, "Look at his feet!"

"Guess they need trimming," the dairy farmer said.

"Trimming! He's foundered from eating silage!" the woman exclaimed.

Charlie is now back at the Standardbred farm, but is a permanent cripple. Even after months of trimming by an experienced farrier, he can only hobble on soft ground. Like far too many other ponies, Charlie Brown had nearly been killed by inexperienced kindness.

The person who coined the phrase "eating enough to kill a horse" didn't know anything about the equine digestive system. The horse has the most delicate and most easily upset digestive system of all domestic animals. If you understand this, however, and follow a few simple rules, you can feed horses for a lifetime and never have a problem.

THE EQUINE DIGESTIVE SYSTEM

The horse's digestive system is made to handle roughage the same as that of a cow, goat, sheep, and other ruminants. The horse, however, is not a ruminant. In the ruminant food goes directly to the rumen, or first stomach, where it is broken down by bacterial digestion before being chewed again as "cud." Then it passes through the second and third stomachs into the fourth, or true, stomach. In the horse, food goes directly to the relatively small stomach and only after passing through the small intestine does it reach the large cecum, or "blind gut," where bacterial digestion does take place. If because of eating too fast or having poor teeth the horse doesn't chew its food well, the food can't be handled properly in the stomach. It may just stay there and ferment, causing gas formation and pain, producing the symptoms known in the horse as colic. If a horse eats fermented feeds, such as silage, the same thing happens. In such cases the horse may try to vomit, and may actually rupture the stomach, causing peritonitis and certain death. If this poorly chewed or improper feed passes on into the small intestine, it may cause irritation there and violent colic. (See Chapter X.) If poorly chewed roughage makes it through the stomach and small intestine, it can still cause problems by clogging the cecum and large intestine, in a condition known as impaction.

Horses have another digestive peculiarity — overeating of grain or fermented feeds, whether they show signs of colic or not, will cause them to founder. Founder, or acute laminitis, is a disease in which the layers of the hoof wall become engorged with blood, causing severe pain and sometimes permanent lameness. This will be discussed in detail in Chapter X, but you should know now that ponies in particular, as well as Quarter horses and horses of draft types, including heavy Morgans, can founder from silage, green grass, or just continual overfeeding of grain. For this reason, turning a horse or pony into lush pasture is not always considered safe.

FEEDING

Hay. Mature, healthy, parasite-free horses can maintain body weight on two pounds of good hay per 100 pounds of live weight. Hay is the only thing they will not overeat and there is no harm in it being given in larger amounts. As work load increases grain is substituted for or added to the hay given.

Hay for horses should be clean and dust- and mold-free. Grass hay such as timothy is fine, but some legumes, such as alfalfa, are more valuable. Clover is acceptable, too, but is apt to be moldy or dusty. Corn silage and haylage are better left for ruminants. Never feed horses hay that is so newly cut that it is still "heating." All changes in a horse's diet, from one batch of hay to another, from one kind of grain to another, should be made gradually. As mentioned earlier, changing abruptly from barn feeding to lush pasture can cause problems.

Grain. A mixture of crushed oats and corn, with some protein supplement and molasses added, giving it the designation of "sweet feed," is more commonly fed than straight oats. Horses under heavy work (four or more hours per day) need oats for energy, but few pleasure horses are worked that hard. Too much corn, particularly ground corn, is not easily digested by horses. The same goes for barley. Although it is fed in many parts of the world, I have seen severe colic and death caused by rations containing finely ground barley (see Chapter X).

In recent years so-called "complete rations" have been on the market. These contain grain and ground alfalfa in a pellet form along with salt and minerals. Theoretically, they supply everything a horse needs except water. For the average pleasure horse owner they are the easiest way to feed, but if no hay is fed your horse will soon be chewing fences and trees or anything else it can reach in search of fiber. Manufacturers recommend feeding 1½ pounds of these complete feeds per day per 100 pounds live weight of the animal. Most pleasure horses weigh about 1,000 pounds, so 15 pounds a day will keep a horse in good shape without the "hay belly" (harmless but not particularly attractive) that results from unlimited feeding of hay. For the average pleasure horse, 7 or 8 pounds of a complete ration daily, split into two feedings, plus good clean coarse grass hay, will be a simple easy way to feed. (Put an 8-quart pail on the kitchen scale and add feed, marking 2, 4, 6, 8, and 10 pounds on the pail with a marking pen.) As work load increases some sweet feed or oats may be mixed with the pelleted ration.

Horses in hard training or doing heavy work are often given a bran mash instead of their regular grain at the evening feeding prior to a day off.

To make a bran mash mix 4 quarts of *wheat* bran (a light fluffy grain) with enough hot water to make a crumbly consistency. After it cools enough for the horse to eat it, sprinkle 1 or 2 ounces of salt (a small handful) on top and feed. Until horses learn to like bran mash you may have to sprinkle some oats or their regular grain on top. Most horses, however, eat it with relish.

Salt. Salt is the only thing besides hay and water that a horse absolutely needs. A salt block should be available year-round, but during hot weather, if you are doing a lot of riding, the horse sweats a lot and will need salt added to the grain, two tablespoonsful at a time once or twice a day. Horses being trained and used for cross-country and eventing need even more salt and perhaps an electrolyte mixture. Your own veterinarian will help you to choose the proper ones, along with minerals or vitamins needed for special reasons.

FEEDING TIPS

Digesting hay causes heat, so in winter feed more hay. A few hard dry ears of corn fed instead of grain during the winter help to keep a horse's teeth in good shape and provide something to chew on. In summer, when you are probably riding more, feed more grain and less hay, although there is no harm in horses having hay in front of them at all times.

It is said that the complete rations may be left in front of horses all the time without fear of them overeating, but I feel most horses will become too fat if they have their free choice. If a horse eats grain too fast, putting a few stones the size of softballs in the manger will slow him down.

Colic and/or founder may also be caused by graining or watering a "hot" horse. When returning from a workout that has caused the horse to sweat you must walk him until the sweating stops. Offer him a little water at a time during this cooling-out process, until he won't take any more than eight swallows every five minutes. When he is cool, and only then, should he be fed grain, but coarse hay may be given at any time. During cool weather you may have to put a light canvas sheet on your horse until he stops sweating (or a heavier blanket during cold windy weather) but be sure to remove it. If you leave a horse blanketed once he is dry he will require protection all winter.

PASTURE

Pasture is fine for growing foals and brood mares, but as the only source of feed for horses that are to be worked even a few hours a week it is greatly overrated. Horses turned out to good pasture in early summer will fatten and look shiny, but the fat put on by grass just doesn't last with hard work. A few hours a day, or overnight, on pasture is good to give horses exercise, and they obviously do enjoy it. But grain and some hay should be available as well if they are to be worked.

Checking for Pasture Hazards. As mentioned earlier, the first lush grass in the spring will founder some horses and most small breed ponies. Other pasture hazards are poisonous plants. Before turning horses out in a new pasture, check with someone in the area who has pastured there before to

POISONOUS PLANTS

make sure it is safe. Even an innocent-looking apple tree can cause problems. Horses love apples, but turn one into a yard or pasture with an apple tree in late summer and you'll have a colicky horse in less than an hour. If the horse is in the pasture from early spring and gets the apples a few at a time they won't hurt him.

Although horses usually know enough not to eat poisonous plants, owners should be aware that under certain conditions most ornamental shrubs are dangerous. The worst of these is the common yew (*Taxus* species), found in the door yards of many homes. Don't tie a horse near a yew or throw yew clippings over the fence where a horse could eat them. A teaspoonful of yew needles will kill an adult horse.

Under certain conditions common red maple (*Acer rubrum*) has been found to be poisonous to horses. To be safe, never let a horse eat leaves from a tree of any sort when the leaves are wilting.

Deadly nightshade (*Atropa belladonna*), a vine-type weed that looks like a cross between the tomato and potato plant with small purple blossoms and red berries in late summer, is found growing in many old barnyards. Horses normally won't eat this weed, but if you put a horse in a yard where there is nothing else to eat he might try it. When you move your new horse to your new yard, check to make sure the only thing growing there is grass.

Lawn clippings should never be fed to horses. Neighbors often feel sorry for a horse in a bare yard and think they are being kind by giving him lawn and shrub clippings. It takes diplomacy to prevent this, but it is too serious a threat to your horse's health and, in fact, his life, to ignore.

I know that up until now I have told of so many dangers to your horse's health that it sounds as though I'm trying to discourage you from owning one. On the contrary, I would rather warn you of the things that can go wrong so that you won't have them happen and become discouraged.

Housing and feeding your horse aren't really difficult if you follow the rules. To sum up, horses can live very well on hay, oats (or complete ration), salt, and water. Changes should be made gradually and then only to feeds known to be safe. Never allow a hot horse unrestricted water or any grain at all.

CHECKLIST FOR FEEDING AND PASTURE

- **Allow your horse (if mature and healthy) to eat at least two pounds of good hay per 100 pounds of live weight, daily.**
- **Make sure the hay is clean, and dust- and mold-free. Never feed horses hay that is newly cut.**
- **Make all changes in a horse's diet gradually. In particular, do not change abruptly from barn feeding to lush pasture.**
- **In winter, feed more hay.**
- **Avoid a grain mixture containing too much ground corn or barley.**
- **In hot weather if you are doing a lot of riding add a couple of tablespoonsful of salt to the grain once or twice a day.**
- **Let a hot horse cool down before you offer it grain or water.**
- **Do not rely solely on pasture to feed a working horse.**
- **Be sure your pasture is free of hazards, including poisonous plants.**

CHAPTER IV

METTLER

FENCING

***Bill Green** thought he had found his dream horse when he first bought Diablo. A big dark bay Thoroughbred, Diablo was tall and strong enough to carry Bill's two hundred pounds over any fence he was pointed at, yet, contrary to his name, of quiet, even disposition.*

Bill found Diablo at a riding stable in early winter. The stable owner had bought him at the track. He had been sold for the reason many racehorses become riding horses — he was just too slow. Green was an executive with an advertising company and although his office was in the city he was able to spend long weekends at his country home with his wife and children. Next to family, horses were the most important thing in the Greens' lives.

In the spring Diablo was moved to the Greens' stable where they kept five other horses, including a driving pony, a show-class Arabian filly, an Arabian brood mare, a half-Thoroughbred hunter, and a Connemara pony hunter. Diablo would make it possible for Bill to hunt with his wife, who rode the hunter, and his daughter, who rode the Connemara, when the fall season started.

Diablo's new home was a typical small village stable — an old barn that once housed horses, a couple of cows, and a flock of chickens, nicely converted to six box stalls with hay storage overhead. An ell off the barn, which had once been a wagon shed, had been converted to a tack room, feed room, and storage area for a two-horse trailer and the pony cart. The area between the two barns, which had been the classic "barnyard" of years gone by, was enclosed with a four-foot wood fence and was used as a paddock. The paddock gate opened to a three-acre pasture once fenced with barbed wire. The top wire was now covered with boards. The top of the fence posts projected four to six inches above the boards.

The first time I saw Diablo was half an hour after he had been turned out in the paddock. He was bleeding from a wound over the front of his right gaskin, and the front of each rear leg, from the stifle to the fetlock, was scraped bare. He was a docile patient and allowed me to stitch the wound with only local anesthesia, and clean up the abrasions. He almost seemed to enjoy the attention he was getting.

Mrs. Green said she had turned Diablo into the paddock after turning the other horses out in the small pasture. "Diablo didn't try to jump the fence, as he easily could have," she said. "He just half jumped, half climbed it and as he dragged his rear quarters over, one board broke and he apparently cut the skin on a wood splinter."

The next day both of Diablo's rear legs were swollen. I knew he needed exercise but was afraid to suggest he be turned out. I gave him a diuretic and recommended he be lunged twice a day. This seemed to work and after ten days he appeared well enough healed to be ridden. In the meantime the paddock area was rebuilt with a board fence six feet high.

Diablo respected the high fence and he was turned out into the paddock daily after the other horses went to the pasture. All went well for a few weeks until hot weather and flies caused a switch — the horses were now outside nights and inside days. At the same time the Green family went away for two days and left the horses in the care of a neighbor.

The neighbor, an elderly man experienced with horses, felt sorry for Diablo, who was kept in the tiny paddock while the other horses ran and enjoyed the green grass in the pasture. The second night he turned Diablo out with the rest of the horses. Instead of eating grass, rolling, or jumping and running as the old man expected, he seemed bewildered and frightened of the other horses. The pony came over to sniff him and he ran away like a frightened dog. The pony, of course, followed and seconds later Diablo was part way over the pasture fence, again half climbing it instead of jumping, as he easily could have. Then, to the old man's horror, Diablo tried to back off the fence and got stuck on the top of a fence post. Another lunge backward and the post snapped, but not before it pierced a hole under his foreleg.

When I arrived the old man had Diablo in cross-ties and was holding a gauze bag full of cotton up under the leg to slow the bleeding. Besides this severe wound he had numerous lacerations from the barbed wire and a few scrapes from the board on top of the fence, but nothing that compared to the hole under his foreleg.

Again, Diablo's good disposition made it possible to clean the wound, tie off one artery, and stop the bleeding. Sutures were put in to close the wound but I knew that nine chances out of ten I'd have to remove them in a few days to drain and irrigate the wound.

The poor neighbor was almost in shock. I knew he was an experienced horseman, having raised and trained draft horses and Standardbred driving horses until retiring.

"I've worked with horses that were crazy and horses that were stupid, but this one isn't either," he said. "It's almost like some part of his brain never developed. Diablo has something missing."

Due to the hole under Diablo's foreleg he could not be lunged for exercise or even turned into the small paddock for over a week. As is typical of many thin-skinned, high-quality horses, particularly Thoroughbreds, poor Diablo swelled in all four legs. Despite good treatment and lots of care from his owner, he was weeks healing. Again, had he not been of such calm disposition, combined with a Thoroughbred's will to survive, I doubt if he would have lived.

Diablo was not ready to be ridden by the August "cubbing" season, but by October he was back in good health and became a fine hunter. Although Bill Green kept and hunted him for many years he could never quite relax with him. As long as Diablo was in a stall, in a small high-fenced paddock or under the control of a human, he was fine, making no mistakes to harm himself or his rider. But make a mistake like leaving a stall door open or a carelessly tied knot on a tie rope and he'd be off like a brainless mechanical toy, bumping into fences or parked cars. Once he even wandered down the busy state highway while brakes squealed and tractor trailers took to the shoulder. The old man was correct: Diablo had something missing.

The neighbor and I had a lot of time to talk and ponder, the night Diablo wounded himself on the fence post. By the time we put him back in his stall, leaving a night-light so he could see, we had agreed on Diablo's problem. Horses raised under confined conditions, never allowed to run at pasture with their mothers and as weanlings and yearlings, have, in a sense, been deprived of one of the stages of development of the normal domestic horse. Horses that learn about fences while still at their mothers' side, whether Thoroughbred or mustang, never forget the lesson. Horses are, after all, a herd animal and must be allowed to mingle with others while growing up. If allowed to run at pasture as weanlings and yearlings they might run into, or even through, a board fence if frightened. Still, barring this type of severe accident, little bumps, bruises, and pricks from barbed wire teach them respect for fences. A horse that is not fence-wise is a disaster waiting to happen. Diablo was one of these.

If I could have a wish fulfilled, no horse would ever be confined to a paddock or pasture unless it was constructed with smooth plank on the inside and to the top of six-foot posts. It should have no square corners, sharp angles, or traps. But knowing what fencing costs, and knowing that some horses will find a way to get hurt even under the best conditions, I must modify this wish. All horses should be raised as part of a herd as nature meant them to be, learning to cope with an environment that is filled with dangers such as narrow gates, low paddock fences, and barbed wire.

Despite criticism of barbed wire by those who consider only board fences correct, I would venture a guess that at least 90 percent of the pastured horses in North America are behind barbed wire. Some people might say that "these are common cold-blooded horses, you can't put a hot-blooded Thoroughbred or Arabian behind a barbed-wire fence." This is like saying Arabians and Thoroughbreds are not as smart as Quarter Horses and Belgians. It is all in the training and handling of the horse, regardless of breed. The fighting spirit bred into the Thoroughbred and Arabian will cause them to destroy themselves if tangled in barbed wire, or any other wire for that matter. Yet their strong sense of survival will alert them to the danger if they are properly acclimated as foals. As one experienced horseman puts it, some horses will get hurt in a solid-walled paddock. But most have sense enough to be kept behind any kind of fence.

So where does this leave you, a prospective or present horse owner, in deciding what to use and how to build a fence? You must compromise, considering what you can afford and what is absolutely necessary for the safety and well-being of your horses. Whether you have a fence built by specialists or build it yourself certain rules must be observed. You might be better off not having a horse if you can't afford to observe the following rules.

FENCING GUIDELINES

1. Paddock area fences should always be of wood.

2. Areas near stable doors, water tubs, narrow lanes, and either side of gates to pasture should always be wood.

3. Gates should be made of wood or

"Texas"-style gate—*not recommended.*

smooth metal, and constructed so a horse's foot can't be put through and caught.

4. The above fences and gates should be four to five and a half feet high, with six feet needed for stallions or high-spirited horses such as yearling colts.

5. Narrow places or corners where a kicking horse can trap another, or where crowding and injury can occur, should be avoided at all costs.

6. Posts should never extend above the top board of a fence, except where corner posts need to be over seven feet tall for bracing purposes.

7. Metal fence posts should never be used for horses.

8. Posts should always be on the outside of a fence.

9. Page wire, or any wire that has holes through which a foot can be forced and trapped, should be avoided.

10. Electric fence should be avoided except under unusual conditions such as at the top of a stallion paddock. If used as a temporary pasture fence in an emergency it must be marked well to make it visible. Horses should be "introduced" to it by leading them near it without touching it.

11. When any wire fence is used it should always be "guitar-string tight."

12. Never use, even for a temporary gate in an emergency, the so-called Texas gate of barbed wire.

PRESSURE-TREATED WOOD

The introduction of reasonably priced pressure-treated wood, which is easy to use and withstands weather and exposure to earth for forty years, has made it possible to build wood fences for horses that are practically permanent. The first attempt to pressure-treat wood in the 1970s led to a product that resisted rot but was, in some cases, poisonous to animals that licked or chewed it. This gave all pressure-treated wood a bad name, and stories persist that deter horse owners from using it for fencing. But now you can buy wood that is pressure-treated with water-borne preservatives, and redried after treatment. With certain precautions this is safe to use.

The basic element in preserving wood is arsenic (chromated copper arsenate is one example). This causes people to fear using wood containing it, particularly since some horses are notorious wood eaters. Arsenic, which prior to the days of antibiotics was a medication commonly used in veterinary medicine, is rapidly eliminated from the horse's system. It is unlike some of our modern chemicals such as some insecticides which accumulate in the body and poison gradually.

Although there is enough arsenic in treated wood theoretically to kill a horse, as one authority put it, "The amount of wood a horse would have to eat to acquire a fatal dose of arsenic would be fatal in itself."

Still, certain precautions are recommended when using treated wood. First, be sure you purchase it from a responsible dealer and that the manufacturer used water-borne preservatives, re-drying after treatment. Second, do not accept it if it shows caking or build-up of preservative on the surface, which should be clean. Third, do not use it for mangers or where it is in contact with animal or human food. Fourth, always wash your hands after working with new wood, and prior to eating, smoking, or handling food. Fifth, in the stalls or paddocks of horses that are serious wood eaters, use a piece of untreated wood, or cover the wood with metal where most chewing occurs, such as the top board on a half door.

Despite claims of safety, and after taking all common-sense precautions, one must always remember that all animals, including humans and horses, have idiosyncracies. It is possible that one horse or one person in a million might be particularly sensitive to one of the elements used in pressure-treated wood, just as they might be to tomatoes or hazelnuts, for example, and have a fatal reaction to it.

UNTREATED WOOD

Untreated wood costs less but will not last. Until the advent of pressure-treated wood rough sawed oak plank and locust posts were the most durable materials you could buy. Oak plank or boards will stand weathering for twenty years or more. Now, however, oak is unobtainable unless you have your own woodland. If this is the case, don't have it cut until you are ready to build, and then build while the wood is still green. Once dry the boards require drilled holes to put nails through.

Oak posts, on the other hand, will not last because they cannot stand exposure to the earth. Locust, which is just as hard, will last. Like oak plank, locust posts should be used "green" since once dry and hard they must be drilled to admit nails or lag bolts. Locust posts are usually used rough, as split or peeled.

It is possible to buy PVC fencing that is said to be safe and makes an attractive fence. At present the main drawback is the cost. Only time can test its durability in sunlight and sub-zero temperatures.

Building Fences

In most areas there are contractors who specialize in building fences for horse farms. Horse farm owners find it is sometimes more economical to have a professional build fences of correct material and design than to do it with help unskilled in fence building and using inadequate equipment.

A well-built horse fence

No commercial horse farm handling expensive horses of unknown background, some of which have never been "turned out," would have anything but the best-built board or special wire horse fence (*see illustration*). However, if you own one to a dozen horses of known background, you can safely compromise both in the type of fence and in who builds it.

Posts should be set six to eight feet apart for a paddock, and eight to ten feet apart for pasture. Using a mechanical auger post-hole digger is the most practical way to set posts for board fencing and, for that matter, any fencing used for horses. Posts sharpened and driven with a post driver are practical for wire fencing in pastures large enough to make wire fencing safe.

Locust or railroad ties used to be the most desirable material for corner posts. However, with the introduction of pressure-treated posts that will last for forty years, they are used less and less. Corner posts should be braced (*see illustration*) so they will stand solidly enough to stretch wire on, or to prevent board fencing from leaning. Setting corner posts in concrete may sound like a good idea, but once placed they are almost impossible to move should you decide to change the configuration of the fence, or if a post breaks off. When corner posts are at least seven feet tall they need not be cut down to the height of the top board. If less than seven feet, however, they should be trimmed to the height of the top board on the fence, as other posts are. On board fence, or even wire fence enclosures, corners should be protected so that they will not trap horses (*see illustration*).

For quiet, fence-respecting horses, two six-inch boards at 46-inch and 28-inch height will suffice, but this is an absolute

Protecting corners to avoid traps

Bracing corner posts

minimum. Preferably, use four or five boards or planks with the top one 48 to 54 inches high, set close enough together so that horses cannot reach through. If wide enough to reach through, set boards far enough apart so a horse won't get its head trapped. Boards should be attached to the inside of the posts for strength and to avoid stifle and other injuries as a horse runs along the fence.

When using pressure-treated wood or, for that matter, any wood for fencing, the use of galvanized, stainless steel, or other rust-proof nails is a good practice. Common nails rust out before the wood rots. On painted fences the rust is unsightly. Drilling pilot holes for nails makes driving them easier and will avoid splitting boards.

All fence materials have certain drawbacks and some horses' tendency to eat wood is a never-ending problem. Creosote on the areas they chew will sometimes prevent this. Many people and some horses are sensitive to creosote, however, and its use can be dangerous. Some states and counties prohibit its sale since it is considered a carcinogenic.

In small paddocks and stalls metal may be used to cover the worst-chewed areas. Products may be purchased that are supposed to keep horses from chewing wood, but they do not always work. Some wood chewing can be prevented or lessened by supplying coarse hay for horses to chew on during boring hours when they are confined to stall or paddock. All in all, wood chewing is a problem with no one good solution, short of keeping wood chewers behind wire.

OTHER TYPES OF FENCING

Recent introduction of high-tensile smooth wire fencing has been touted as ideal for horses. I have not had enough personal experience with it, but since it lacks visibility it can be used only for quiet, fence-wise horses. Horses that reach and push against fences would not be kept behind it unless it was strung six inches or less apart. Horses

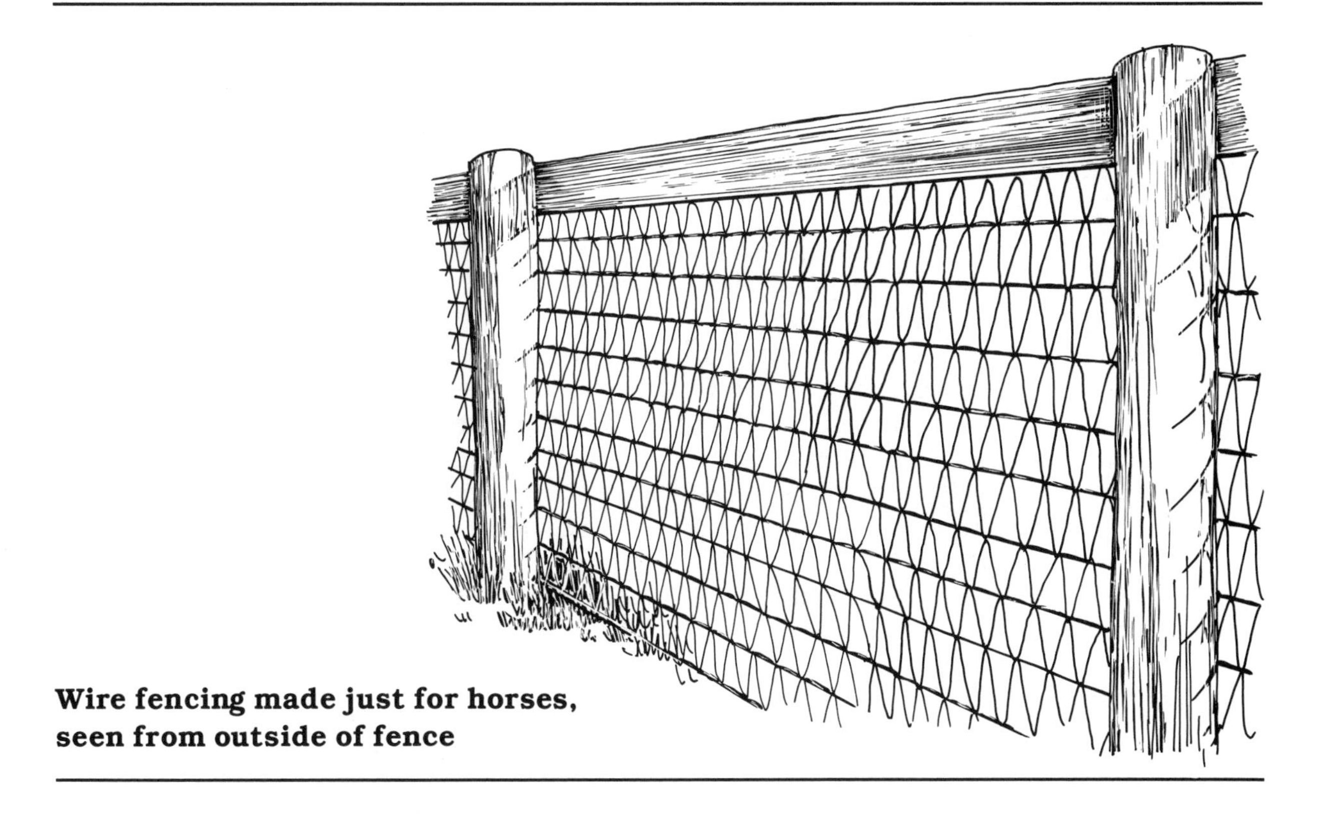

Wire fencing made just for horses, seen from outside of fence

of this sort respect barbed wire better than anything else.

Wire fencing that is made just for horses, with holes too small to put their feet through (*see illustration*) is also available, but very expensive and not easily installed by the inexperienced.

Rubber-covered nylon strips were once thought to be the answer to the horse fencing problem. It worked fine, except horses chewed the rubber off, ate the nylon, and died months later of obstruction to the intestinal tract from the balled-up nylon. (People sometimes put old tires around feed tubs. Tires contain nylon, too. If horses start to chew on them, get rid of the tires immediately.)

BUILDING THE FENCE YOURSELF

If you are building a fence yourself you want it to be attractive as well as practical. Straight lines always look better, but if, for example, you have each top board 48 inches from the ground, the boards will follow the contour of the ground, appearing to go up and down. To give your fence a better appearance, use string pulled tight from corner to corner on nearly level ground, or every four posts on rolling ground to determine the height to place boards or wire.

One is often tempted to put the boards or planks on the outside of the posts along a road or driveway for appearance's sake and then paint the fence white. If you do this and are going to keep horses behind the fence, be sure to put at least two additional boards on the inside to prevent horses from pushing outside boards off (*see illustration*) or, worse yet, injuring themselves as they run along the fence.

Lead paint is almost unavailable now, but be sure never to use it anywhere that animals can lick or chew it.

Books and bulletins are available (see Garden Way Publishing bulletin "The Best Fences" #A-92) on the mechanics of fence building. You can learn from experience, and if you are of a mechanical mind and take time to observe others' accomplishments and mistakes, you can build a good fence. The most common mistakes are setting posts too far apart and not deep enough, and not properly bracing corners.

If you are inexperienced, before starting a fence ask an experienced horse owner to look at your situation and your plans, for safety and practical aspects. Fences are quite permanent and it is easier to change your plans than to change a fence.

SAFETY FIRST

For the safety of the builder of fence, be careful with mechanical devices such as augers and post drivers. Never work with wire of any sort without good heavy leather gloves. If you are driving pole barn nails be sure you have a hammer that won't chip, and remember that some pole barn nails will chip. Goggles or shatter-proof glasses should be worn when driving these hard nails.

Turning the Horse Out

When you turn a new horse into a fenced area for the first time, take him in on the halter and lead him around the inside border before turning him loose. If you have used wire, particularly high-tensile or barbed, make it easily visible by putting little strips of white cloth every few feet on the top wire. If you change a fence for any reason and put up new wire, do the same thing. Horses depend on their eyes to warn them of danger.

Never turn a horse out alone in a new pasture if he is used to being with another horse. Never remove one and leave another alone until you have made sure the loner won't charge the fence in desperation to get out.

If you have a horse that has been at the track or show ring for months and has not been turned out, take precautions the first time, even in a paddock. Certainly use a board fence paddock for first turnout before turning the horse into a large pasture. Walk the horse around the inside of the paddock fence first. Then, if at all possible, have two or more helpers to stand in corners. If you have only two assistants, put them at the ends of the paddock to wave the horse off if he panics and runs wildly. I have seen horses do this several times and it is one of the most frightening, helpless feelings there is. Once a horse is loose and running, even the most placid may become like a wild animal. Usually the panic passes over in minutes, and if you have enough help to wave him off, a sensible horse soon settles down.

Another method of turning out for the first time is to use a small dose of tranquilizer. Remember, however, that some tranquilizers seem to affect horses' eyesight and common sense so, if possible, have help with you at the first turn-out of a strange horse. Chances are you will never have a problem, but avoiding one is far easier than stopping it once it starts.

There is a certain feeling of self-satisfaction in building a good fence. Both the bank mortgage inspector and casual passerby judge an owner by the quality and appearance of fencing. A little extra time taken first to plan fencing and then to build it properly can add much to your enjoyment as a horse owner.

CHECKLIST FOR TURNING A HORSE INTO A NEW PASTURE

- **Lead the horse around the fenced area before turning it loose.**
- **Never turn a horse out alone in a new pasture if it is used to being with another horse.**
- **Have strips of cloth tied to the top wire of the fence to make it easily visible.**

Posts built correctly on outside of fence

CHAPTER V

AMERICAN QUARTER HORSE ASSOCIATION

HANDLING YOUR HORSE

***Amanda** was the daughter of a client who had dairy cattle for a business and trained Standardbred horses as a hobby. When she was eight years old her father found Copper, a red chestnut quarter-horse type with unusual blonde mane and tail, at an auction and brought him home. Amanda was soon riding him not just on the farm and around the training track, but all over the area, usually bareback and at a gallop. The pretty little girl, blonde hair flying in the wind, seemingly part of the red horse with its flying blonde mane and tail, reminded one of a young Greek goddess. When anyone spoke of caution or the danger involved, the answer was, "Copper actually looks after Amanda. You couldn't make him do a bad thing like kick or strike, and riding bareback there is less chance of her getting hurt."*

After a few years no one worried about this any more. Amanda was twelve, riding with an English jump saddle, wearing a safe hard hat and boots, jumping, showing, and still a familiar sight on local roads. One windy day Amanda and a friend rode into another friend's yard across from the town parking lot and dismounted. For some unknown reason Copper and the other horse left the grass they were eating in the yard and crossed over to the parking lot where Copper wandered between two cars. Amanda ran across the street, calling to Copper, and grabbed his tail to drive him out from between the cars. Whether Copper did not hear her due to the wind, or was already nervous from the wind and the strange surroundings, will never be known. He kicked straight back and hit Amanda full in the face with a freshly shod rear hoof.

For days Amanda's life hung in the balance. The plucky little girl survived, but had to suffer through months of surgery to repair her face and mouth. There is no such thing as the horse that will never kick.

To **work** with horses one must know a little of their unique personality or psychology as well as their anatomy and physiology. The original wild horse, ancestor of the modern horse, was an animal that depended on flight for survival, just as deer and antelope do. Like all animals hunted by beasts of prey, the horse's eyes permit it to see to the side instead of straight ahead.

The horse is by nature a nervous, flighty animal, yet it can calm down and relax as quickly as it gets excited. In the beginning all horses were similar, but down through history man has bred them to develop certain traits, thereby creating different breeds. The various breeds differ not only in conformation but also in abilities and temperament. In general the heavy draft horse is less high strung, for example, than is the racehorse.

We humans judge intelligence of animals by our standards—how large a vocabulary a gorilla understands versus how many words a horse, dog, or cow understands. Sometimes we judge an animal's intelligence by the number of "tricks" it can be taught to do. By those standards a horse is as smart as anything short of the great apes and perhaps the dog. On the other hand, some people say if a horse were truly smart, it wouldn't let man make it work so hard.

As a veterinarian working with horses, cattle, dogs, and cats I am often struck by the similarities between dogs and cows and between horses and cats. One can bully a dog or cow and force it to submit. When it comes to horses and cats, however, the human cannot force them, but must "talk" them into wanting to do what is expected of them by touch, eye contact, and voice.

The rest of this chapter will be devoted to telling you a few of the things that will enable you to handle an animal that is ten times as heavy and ten times as strong as you. The more you work with horses the more you will learn about them, and the more you will be awed by the wonderful relationship between horses and humans.

Growing up on a farm where horses were the main source of horsepower, where my older sisters drove a horse and buggy to the local high school, I learned certain things that serve me as instinct or reflex. If these are not learned by a new horse owner they can lead to disaster, or at the least, lessen the joy in owning a horse.

The first thing is to say "whoa" or just plain "ho" every time you approach a horse, to make sure the horse knows you are there.

Never surprise a horse or approach it silently. Even while you are grooming or tacking a horse, if you step away or lose touch contact for an instant, or even change your contact from one spot to another, say "ho." I learned this as long ago as I can remember from my father. A gentle man, he would never strike a horse or cow, but would remind me with a cuff if I didn't speak to an animal when I was moving within kicking distance. Like most of his generation he knew from sad experience that the most gentle horse in the world can kill you with one kick if startled.

Horses' eyes are set on the sides of their heads so they can neither see clearly directly in front of them, nor directly behind them. Therefore, always approach horses from a bit of an angle, speaking to them as you do. The one-eyed horse is doubly dangerous since it is shy on the blind side. Horses are trained to have an off side (right) and near side (left). If possible always approach from the left side, and be doubly careful if you must approach from the right. This is imperative when entering a stall, particularly a straight stall.

If one is afraid of horses it is difficult, although not impossible, to overcome it. But no one, no matter how unafraid, should ever lose respect for the fact that a horse can seriously injure you by kicking, striking (using a front foot to reach up, drive out and down), biting, or just stepping on you. Never work with a horse when wearing sneakers, sandals, or in bare feet; always wear hard-toed boots or shoes.

Children, in particular, should not be allowed to become so unconcerned with horses that they wander under them, walk up behind them unnoticed, or rub their faces against a horse's nose or mouth. One of the worst things I've seen in recent years was a magazine cover photo showing a horse nuzzling the top of a child's head. Children and adults seeing such a photo could believe that it would be proper and safe to let any horse do this. Only weeks before I saw this magazine cover, despite posted warnings to "keep out," a woman held her two-year-old child up to the half-door of a stall and the horse bit and seriously damaged the child's face.

GOOD DAYS AND BAD DAYS

Just like you and me, horses have good days and bad days. It is not superstition that horses are "cranky" on the first cool frosty morning of early fall, or "spooky" on windy days.

One of the first horses I knew was Tom, a medium-sized Roman-nosed work horse that, I believe, loved my father as much as my father loved him. Tom would work all day beside his teammate Jerry, pulling a mowing machine in one of the high back lots of the farm. As the twelve-year-old "driver," I had little to do but raise the cutting bar when we came to a rock, or as the horses took a square turn at each corner. The noise of the old Walter A. Wood mower was louder than any modern tractor and I could not hear the wind in the trees toward the mountain or the call of the crows, nor sounds from the valley floor half a mile away. Yet suddenly in late afternoon Jerry would start to turn at the lower end of the field and Tom would refuse. Stopping the mower stopped the sound, and far away I could hear the "Yoo-hoo" from my father, letting me know it was time to come home for chores. Tom had heard it above the racket of the mower. To this day I believe that old Tom and my father had mental telepathy, a sense of communication between them.

Yet in September each year on the morning of the first frost, when Dad went in to harness Tom, either as he placed the collar or as he tightened the hame strap, Tom would reach down and bite him on the arm, sometimes hard enough to draw blood. Dad's response was always, "Tom, why did you do that?" Anyone else I've ever known would have slapped the grouchy old gelding on the nose and he wouldn't have done it again.

HOW TO TURN A HORSE OUT

If you *are* afraid of horses, the more you learn about them and the more you work around them, the more confidence you will develop and your fear will be replaced by respect. There are many little things one must do to practice proper safety with horses. When you walk a horse to a pasture or paddock or, for that matter, return it to its box stall, try the following procedure:

1. Walk on the horse's left, even with its head. Never lead a horse with just a halter. Always use a lead shank or rope.

2. Hold the shank or lead rope short so you don't get crowded against the gatepost.

3. Walk all the way in until the horse has cleared the gate.

4. Stop and make the horse stop.

5. Turn the horse to the left slowly until the horse is facing the gate and you are facing the horse. Make the horse pay attention to you —not just awaiting being freed.

6. Back up, facing the horse, until you can reach the gate and partially close it with one hand.

7. Step backwards through the gate (or stall door), and say "whoa" to the horse.

8. Either remove the halter or unsnap the shank or rope, step back behind the gate, stall door or gatepost, and close it.

This may sound like a lot of military-type procedure that is done to waste time. Like military procedure, however, it is done for safety. If you just step into the stall or paddock and let the horse go it can, with no mean intent but just the joy of sudden freedom, run past and kick you with the near rear hoof. Or the horse may decide it doesn't want to be in the stall, and rush out past you. Once you make the above procedure a routine, done by reflex, others who notice it will know you are a real horse person.

Horses as well as people often hurt the one they love most.

Develop your caution of horses until by reflex you say "ho" without thinking. But never forget that even the horse you love the best and the one that returns your love can have a bad moment and could kill you.

Learn Basic Safety

CATCHING A HORSE

In Chapter II we discussed leaving halters on or off when horses are turned out to pasture or in a paddock. For the first few days with a new horse, until you are sure you can catch it, it is best to leave the halter on. Horses usually learn to come when called if they know you are going to grain them. To be sure they come, whenever you catch them, at least have a carrot, apple, handful of grain, or some treat — even a lump of sugar or bit of chewing tobacco — to reward them.

Most well-trained horses will put their noses in the halter nose loop when you hold it out. Some can be absolute "brats" about being caught unless they have a halter on, even in the stall. One way to overcome this head shyness is to pass a lead shank under and over the neck. Many Western-trained horses will give up immediately with this and then you can slip the halter on.

With others you must begin to stroke the neck as they eat grain from a pail. It took me years to realize that stroking a horse is what they like, *not patting*. (One may pat a horse to say "well done" after a jump, but that is a different situation.) Stroking has a calming effect on horses. Start at the neck if you are at that end, or if trying to catch a horse from behind, stand at a 45-degree angle to its left rear quarter and stroke its buttock, gradually working up along the back to the neck as you feel the horse relax. Then as you stroke the left side of the neck with one hand either work your other hand, carrying a bale string or soft lead rope, under and up over the right side, or if the horse has a halter on, reach and take the halter gently. Never make a quick move.

HORSE TALK

During this entire sequence talk softly or in a firm commanding tone. I can't tell you which is better; if you have the instinct you will know by what the horse says to you. Horses do not actually speak in words, of course, but, like all animals and birds, communicate in subtle ways, as we humans apparently do to them.

Another way to approach certain horses, particularly those that are headshy, is to walk into the stall, say whoa, and walk directly to the left front leg and pick it up. Then after a moment reach up and take hold of the halter. No matter how easy it is for you to catch your horse, it probably won't let the veterinarian or farrier catch it, or even let you catch it when the veterinarian's or farrier's vehicle is in the area. So remember, when you are expecting either of these people, have your horse confined before they come.

While working around a horse you are not familiar with, watch its eyes and ears. They are the most evident of its subtle ways of communication. Eye contact with horses is just as important as eye contact when dealing with other humans. Some horses respond to gentle blowing from your nose or mouth, and some are calmed or frightened by odors. Farriers tell me they prefer to wear an old leather apron when working around a skittish horse. The smell of other horses apparently has a calming effect. A veterinarian must wear immaculately clean coveralls when working on horses. Remembering the farrier's advice, however, I always like to wear a well-worn leather cap that can't help but smell like horses.

Some people can smell fear in a horse.

Although I've never been sure of it myself, I know and am only too familiar with the fear smell in cattle, best described as the odor of sweet fern. Horses certainly must smell fear in us and in each other. Although it is seldom mentioned, some normally well-behaved horses, particularly stallions, will become unruly or violent when handled by a woman during her monthly menstrual period. This is not a superstition but a fact, and not to mention it here would be negligent.

RESTRAINING A HORSE

Before you can give a horse first aid or, in fact, before you can even have the veterinarian or farrier carry out routine care, you must be able to handle the horse in a calming way. Horses feel most secure in their own stalls and will often stand there better for grooming, tacking, and examination. However, light is often poor in stalls, and there is more potential danger to the handler and

Cross-tying a horse

TWITCHES

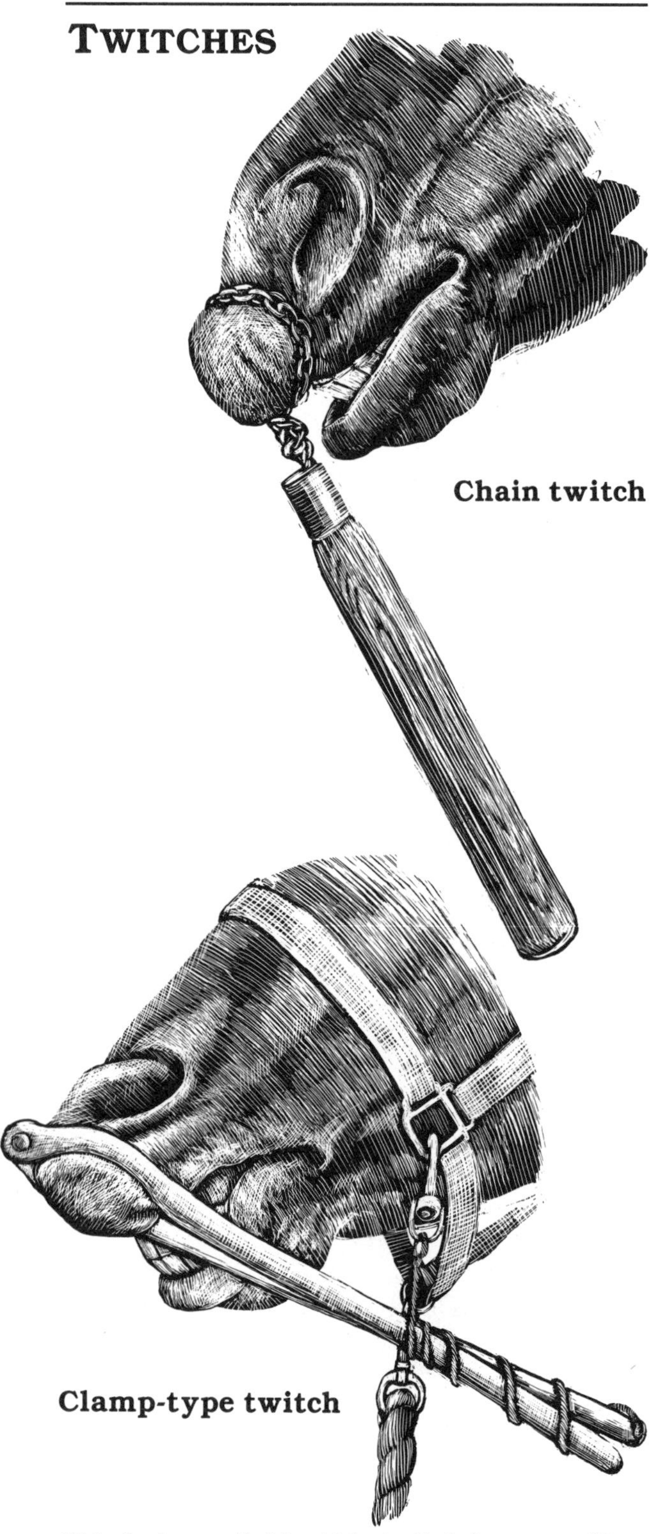

If help is available, this twitch is more effective when held by hand and the cord is *not* used.

others because of the confined space.

The best place to work on horses, safest for the horse and the people concerned, is in cross-ties. Cross-ties are simply ropes or light chains with snaps that reach across an open space, almost but not quite reaching each other. The floor in the cross-tie area should be plank, dirt, or, if concrete, covered with a heavy rubber mat. Cross-ties may be used in a box stall, but the horse should also be trained to stand in cross-ties in an alleyway or doorway where there is room for the handler to move safely should a horse kick. To secure the horse the snaps are hooked onto the side ring of the halter or, in some cases, to the tie ring itself. Once a horse is properly broken to cross-ties a person alone can groom, tack, work on feet, give medical treatment, and carry out minor procedures such as dosing or taking blood samples with no further restraint.

If more restraint is needed, release one tie, usually the left or near side, and substitute a chain shank either over the nose or, in some cases, through the mouth as you would a bit. Some people put the shank under the upper lip and over the gum. This is a severe restraint and usually works, but if a horse does rear it can cause serious injury to its gum and upper lip. Well-trained horses will stand for some fairly painful procedures such as intermuscular injection with no more than a chain shank hooked to the halter and the calming effect of one hand over the bridge of the nose. For procedures such as this it is safer to release both cross-ties so that if the horse "flies back" the halter won't break.

Further restraint may be obtained with the use of a "humane" clamp-type twitch on the nose. Recent research has shown that the use of the twitch has a calming effect similar to acupuncture. The use of the chain twitch, which if improperly used can cause the nose to go numb and thus lose its effect at the time it is needed most, should be left up to the veterinarian or other professional horse person. On occasion, perhaps to allow the veterinarian to give intermuscular injection, the use of the "hand twitch," simply grasping the horse by the nose and squeezing gently, will calm it sufficiently.

Although grabbing an ear or ear and nose sometimes works on common horses, it nevertheless always appears to be a crude, cruel method of control. Other methods of restraint, all of which have their place if used by skilled people in a humane way, are: the so-called "war bridle," simply a light cord noose over the back of the poll and down under the upper lip; twitching a lower lip; picking up one leg; grabbing a handful of hide and squeezing or grabbing a handful of mane just ahead of the withers.

Judicious use of tranquilizer before a horse "blows up" and the use of ropes to confine or control horses are fine in their place, but best used by experienced horse people such as trainers and veterinarians. Probably 99 percent of the time a little patience and horse sense on your part can accomplish more than all the mechanical and chemical restraints there are.

Picking up a horse's foot. Hold the horse's back leg over your knee to examine its hind foot.

HANDLING THE HORSE'S FEET

We have mentioned picking up a horse's feet but have not described how to do it. To pick up a front foot, speak to the horse and stand at its shoulder facing the rear. Slide one hand down along the front of the horse's leg to just below the fetlock. Then grasp the pastern and push your thumb in just under and behind the fetlock, pulling back and up just as you push your hip in toward the horse. At that instant you can quietly command the horse to "pick it up" or "let me have it." If the horse is well trained it will work easily; if not, it will be a struggle.

To pick up a rear leg, speak to the horse, standing facing back, in line with its hip. Speak again and push the hip bone with the hand closest to the horse. Slide the other hand down along the front of the left until just below the fetlock. Grasp the pastern, push your thumb in under the fetlock and pull back and up, placing the leg over your knee.

If a draft horse is too large to grasp around the pastern bone, grasp a handful of fetlock hair.

The more time you spend working around horses the more comfortable you will be with them, and the more they will trust you or, one might say, the more communication there will be between you. Like people, you may find certain individuals with whom you just can't communicate. It is a rare human who can get along equally well with every horse. Conversely, everyone deserves one perfect horse in their lifetime. For my father it was Tom, for me it was a one-eyed pony named Nero. You have one, too; just keep looking.

CHECKLIST FOR HANDLING A HORSE

- **Always say "whoa" or "ho" every time you approach a horse, even if you have been away only for a moment.**
- **Always approach a horse from an angle, from the left if possible.**
- **Always wear hard-toed shoes or boots around horses.**
- **Do not allow children to wander casually around horses, or to nuzzle their faces together.**
- **Leave the halter on a new horse for the first few days until you are sure you can catch it. After that, let it be in the pasture or stall without a halter.**
- **Confine your horse *before* the veterinarian or farrier comes.**
- **Learn how to restrain your horse when necessary.**

CHAPTER VI

BONNIE LINDSEY/HORSEMAN MAGAZINE

GROOMING AND FOOT CARE

*V**elvet** was a dream horse, shiny black with a white star and one white sock, gentle of disposition yet full of life. I first saw her at a riding school at a summer camp. The trainer who ran the school was, by well-deserved reputation, "the best." She supplied horses and instruction for young riders at a typical summer camp for city children from well-to-do families. Her horses, too, were the best for that purpose and with this combination she was able to teach children who had never touched a horse the basics of safe, enjoyable riding during their few weeks of summer camp.*

Camp horses ordinarily lead a rough life, the more suitable ones being ridden from daylight until dark. By summer's end they are bone thin. However, this trainer had all good horses and took excellent care of them. Velvet was the most popular of the horses at the camp not only because she was really black (most so-called black horses have a bit of brown on their muzzles and are technically "dark bay or brown") but also because she loved children. She would walk slowly for a young inexperienced rider, but trot out smoothly and canter softly as the rider gained confidence.

The trainer told me Velvet had been her own personal horse when she was a girl, and that year was close to twenty-five years old. As the summer wore on Velvet began to show her age, becoming thin, moving stiffly, and losing the well-earned "Velvet" appearance of her once-shiny black coat. September was coming and we hoped that rest and pasture would return her to her usual good health. This would be her last year as a camp horse.

On a routine call to the trainer's stable in October I was told Velvet had not responded to "turning out," but was even thinner and walked as though she were arthritic. We wormed her, checked her teeth, which were in excellent condition for her age, and suggested an increase in her grain ration. By November, with her new winter coat, she looked better, but still moved as though she had arthritis.

In the spring Velvet still appeared thin, but more important, she seemed more stiff and sore than ever. The trainer had decided to put her down, feeling, as most of us do, that after a good faithful life, if a horse doesn't respond to being turned out to pasture and seems to be suffering, the most humane thing one can do is to painlessly end its life.

The local farrier was at the stable that day, so he and I discussed the old mare. He picked up her feet and showed me that they were short, thin-soled and tender. Like most horses turned out to pasture in the early fall, Velvet had had her shoes removed. Usually by spring horses' feet are long but healthy from running in the mud and snow. This had been a dry cold winter with little snow or mud, leaving bare the frozen ground that wears feet down.

I knew that Velvet, having unusually good teeth for a mare her age, was able to eat normally. The farrier suggested that for years Velvet had been the boss of the herd, but that now, due to her sore feet and advancing age, the other horses drove her away from the hay rack. She was suffering from malnutrition, because she simply couldn't compete for her feed.

Knowing that my children wanted a horse, I asked the trainer to sell Velvet to me instead of putting her down. That night she telephoned me, not calling to put Velvet down, but to tell me that although she would not sell the old horse she would give her to me. I had to promise, however, that if Velvet didn't respond to being alone and properly shod I would put her down.

The trainer trailered Velvet to us the next day. The old mare had her own two-acre pasture, her own stall, and five children who loved her. The farrier shod her the following morning.

That fall the kids showed her at a local show. The old judge placed her first in both the fitting and conformation classes, saying, "She looks just like a mare I used to see at the shows twenty years ago. Would she be a filly out of that one?" The farrier, standing by the rail, just beamed, knowing his shoeing had made the difference.

Before you ever buy a horse start inquiring about local farriers (horse shoers or blacksmiths). Many farriers now specialize in certain types of horses — hunter, jumper, runner, gaited, draft, pacer, and trotter. For your first horse you want a capable farrier who is willing and able to shoe a just-plain riding or driving horse, and who enjoys teaching you about hoof care.

There are entire books on shoeing and corrective shoeing (to correct a fault or lameness), but all you need to know is how to choose a good farrier. Your veterinarian or local horse owners can suggest some to try, but you must make the final decision.

Briefly, you should know that horses' hooves grow fastest while the days are getting longer, and slower when the days are shorter. In spring they may need to be trimmed every four weeks; by late fall that time interval may be doubled. Horses used only on sod and soft ground may not need to be shod at all or, Western-style, shod only in front. (There are two reasons for this: a horse carries about 60 percent of its weight on the front feet, and some range horses won't let you pick up their hind feet.) White hooves are not as durable as black or dark hooves. The striped hoof of the Appaloosa is said to be tougher than any, but I cannot say from experience whether or not this is correct.

You will learn that when shoes need replacing the nails show up through the hoof wall higher than when first placed. It is better to have the farrier come on schedule than to wait and call when you need trimming or re-shoeing. During the busy season the farrier may be scheduled so far ahead by the time you call that you will be in trouble.

PARTS OF THE HOOF

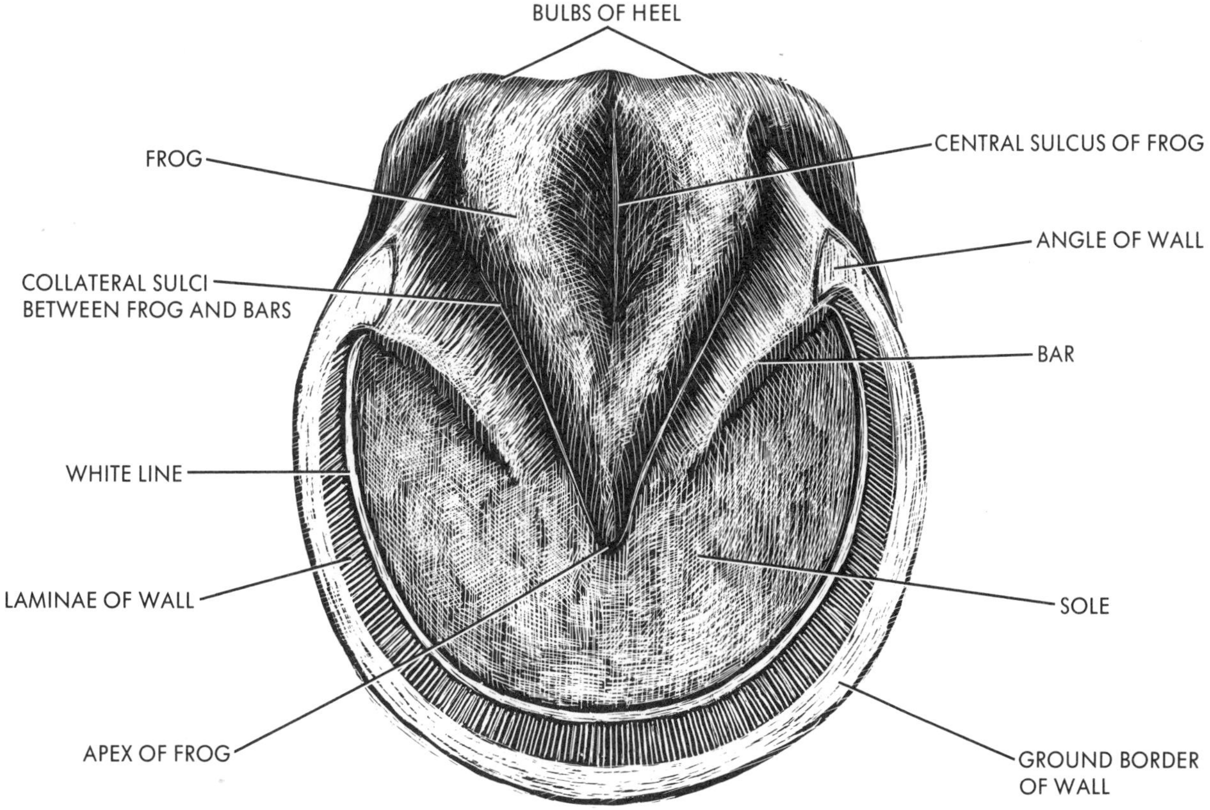

When picking the hoof, work from frog toward toe.

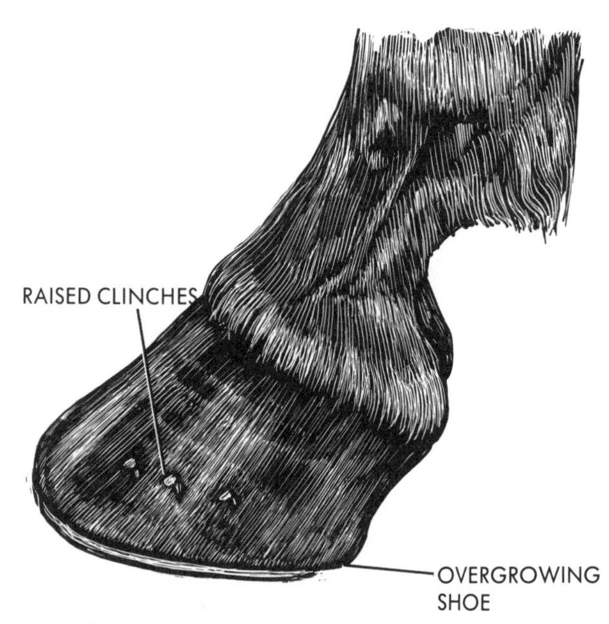

Hoof's appearance when shoe needs replacing

Care of the Hooves

You should buy a hoof pick when you buy your first horse, and make it a habit to pick out the accumulated debris from each hoof prior to and after each ride. Folding hoof picks are easily carried to remove sticks and stones that may be picked up during a ride. To use a hoof pick, simply pick up the foot as shown at left and dig packed dirt and debris out of the sole of the hoof and from around the frog.

If a horse suddenly starts to limp during a ride or drive, the foot on the leg that appears lame should be checked immediately for a stone or, worse yet, a nail. If you find a nail pull it out and as soon as possible pour tincture of iodine into the hole. Call your veterinarian if the nail was in deep enough to draw blood. After having a nail removed a horse may not appear lame for several days, and then may suddenly go dead lame. For this reason, always memorize or, better yet, make a diagram of where the nail was in the hoof so the original hole can be found for treatment.

On days when you do not ride it is a good idea to check horses' feet at least once for stones and other accumulation. During the winter, when horses are best left barefoot, check for feet being too long and starting to split or crack. If this is found call your farrier for trimming before the horse goes lame.

EXTRAS

Under ordinary circumstances hoof dressings are not just a waste of money, but they actually do more harm than good. Most of them block the normal absorption of water that keeps the hoof soft and pliable. Dressings are used for cosmetic reasons on show horses, but once you start you must continue with them. Racehorse trainers use hoof packings (not to be confused with dressings), which add water to the hooves of horses that stand in dry stalls. For the average horse that is outside in a paddock or pasture at least part of each day the hoof absorbs enough moisture to keep it healthy. During a late summer dry spell it might be necessary to create a mud puddle for horses to stand in for fifteen to thirty minutes a day to moisten their hooves.

Pads. Some farriers put full pads of leather or plastic material between a horse's foot and the shoe to protect the sole from stone bruises and to reduce shock. These cause more problems for ordinary riding horses than they prevent. Either hooves dry out too much, or moisture gets between the sole and pad and causes an infection called thrush. In certain cases, however, pads are the only way to keep some horses sound. In the case of horses driven on hard pavement, pads are a must.

Grooming

Horses should be groomed daily, although most are not. A few minutes' once-over with a dandy brush removes loose dirt and seems to make a horse feel good. Never use a metal curry comb on a horse. Even the rubber curry and the stiff brush should not be used over bony areas such as the face or lower legs.

Brushing with the dandy and then a soft brush should be routine before working a horse. Horses should be brushed again after work. Except in the hottest weather, most horses ridden for pleasure will not be so sweaty that they require washing. A good brushing is all that is needed.

After a long ride or drive, when horses

Grooming diagram—direction of hair growth

are sweaty, walk them either by leading or riding until they are dry or nearly dry. Do not let them stand in a cold wind while you have lunch on a trail ride. If you can't keep them walking, tie them in a place protected from the wind and blanket them. In Chapter X (Colic, Choke, and Founder) we shall discuss further the cooling of horses with more detail about water and feed.

On very hot days wash wet areas down with plain water that is 75° or warmer and contains no soap or detergent, since daily use of these removes too much natural oil. Then use a sweat scraper to remove the bulk of the water, and apply a cooler, sheet and/or blanket (see Chapter X) until the horse dries. Once a horse is dry, remove blankets and sheets so the animal will not habitually require them.

In the spring when horses shed their winter coats extra grooming is necessary. Judicious washing on a warm day, followed by a good scraping with a toothed sweat scraper or shedding blade, can cut the shedding time to a minimum.

A clean, well-shod horse not only looks good, but makes both horse and owner feel good.

CHECKLIST FOR GROOMING AND FOOT CARE

HOOF CARE

- Choose a good farrier.
- Trim your horse's hooves more often when the days are getting longer, less often when the days are getting shorter.
- Schedule the farrier's visit ahead of time; do not wait until the last minute.
- Use the hoof pick on your horse's hooves before and after each ride.
- If your horse starts to limp during a ride or a drive, examine the hoof immediately. If you find a stick, stone, or nail, remove it. If a nail is in deep enough to draw blood, call your veterinarian.
- Check hooves daily even when not riding.
- In winter, check that feet are not too long, or starting to crack. If so, call the farrier.
- Avoid hoof dressings if possible.
- If you drive your horse on hard pavement use hoof pads.

DAILY GROOMING

- Brush your horse daily with a dandy and then a soft brush.
- Never use a metal curry comb on a horse, and use only a soft brush on bony areas such as the face and lower legs.

AFTER A RIDE

- Brush the horse before and after the workout with a dandy and a soft brush.
- After a long ride or drive, walk the horse until it is nearly dry.
- Wash wet areas with plain warm water, use a sweat scraper, and then cover the horse until it is dry.
- Once your horse is cool and dry, remove the blanket or sheet.
- Never make a horse stand in a cool wind when it is hot and sweaty.

CHAPTER VII

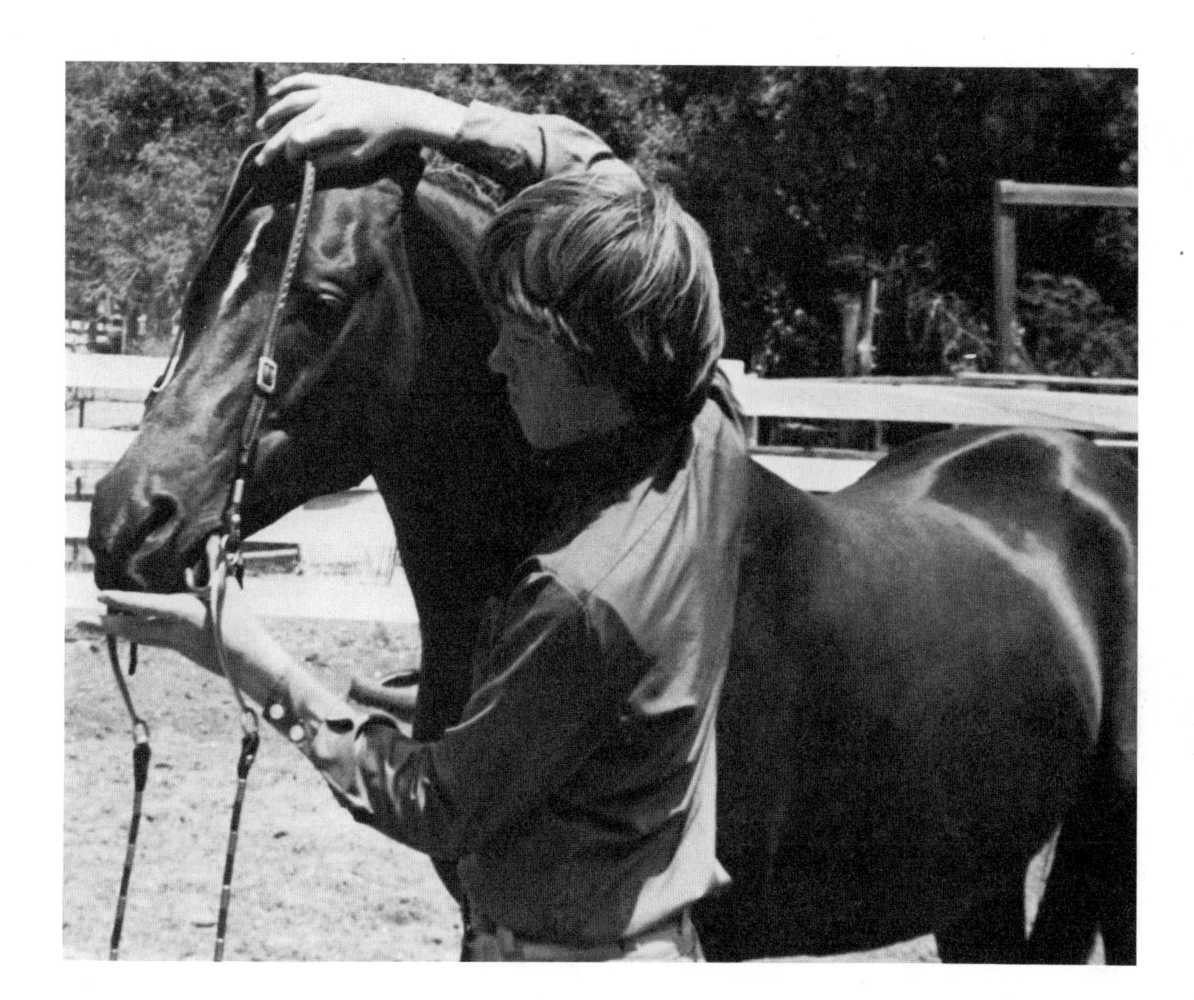

INTERNATIONAL ARABIAN HORSE ASSOCIATION

TACK

One ***of the pleasures*** *of being a veterinarian working with horses is acting as veterinary judge at competitive trail rides. My first experience at this duty was in the early 1950s when the sport was new to both the judges and the participants. Consequently, the participants, the lay judge (who judged riding ability and horsemanship), the timers, the officials, and the veterinary judge learned from and taught each other.*

Judging one or two rides a year, including both adult and junior riders, gave me the opportunity to see some of the same individuals, riders and horses, year after year. Recently a woman who has been a riding instructor, a 4-H leader, and a lay judge over the years asked me what I thought was the greatest change in trail riding from the 1950s to the 1980s. My answer: "The improvement in the quality of horses and tack." She agreed, and then asked, "Do you remember the Swensen kids? They were the best example of improvement, from the bottom to the top, in trail riding and horse show circles."

I remembered the Swensens well, five girls ranging in age from nine to fourteen. Both their mounts and equipment were a castoff collection of all description. The nine-year-old rode a barefoot mixed-breed pony with a red donkey saddle and a bridle that looked as though it belonged on a draft horse. One of the girls had a skinny old Saddlebred with eyes that appeared ready to pop and go into orbit, a huge Western stock saddle, and an English bridle. The eldest girl, mature for her age, rode a little dark bay Mustang using a McClellan saddle and a hackamore for a bridle. None of the Swensens' mounts had saddle pads or blankets to protect them from the poorly fitted saddles they carried.

The kids themselves wore English, Western, and jodphur boots, or sneakers. They wore everything from jodphurs to shorts, with shirts but no jackets or hats.

Despite appearing as a small mounted revolutionary band, the girls themselves were the most polite, good natured, enthusiastic, and likable kids one could find anywhere. On their first trail ride I wanted to disqualify the barefoot pony before it began to go lame on the hard gravel road. The lay judge disagreed, saying that she knew these kids rode miles every day and she had never seen the pony shod or lame. At the end of the 35-mile ride on that hot, humid July day the pony checked out in the best shape of any mount in the ride. His pulse and respiration were exactly the same as at the pre-ride check, and his feet were as cool as when he started.

The other four Swensen mounts also appeared to be in good shape at ride's end, except that two of them had tender spots where ill-fitting saddles had sat. One had a girth sore and the little Mustang had a raw spot the size of a silver dollar on each wither where the old McClellan saddle had rubbed.

The five Swensens listened carefully to criticism from the lay judge and me, and seemed to take seriously the challenge to learn to improve their tack and horse care. During the following year they participated in every clinic and learning session that the county 4-H offered, and two of the older girls joined Pony Club, where they could learn more.

A year later I saw them at a horse show. They rode the same horses but somehow they looked better. They still had the same saddles, but all had pads or blankets, and the girls all wore proper boots and either hard hats or proper Western hats.

Three years later I had the opportunity to judge at a regional 4-H Presentation Day, where children give a five-minute demonstration complete with charts and props to show certain skills and procedures involved in their particular 4-H project. The highest-scoring individual, who went on to state competition and the championship, was Karen Swensen. Her subject was "The Selection of Proper Tack For Your Horse."

In this book I'm trying to tell you the things a veterinarian with a horse background knows about getting along with and enjoying horses. When it comes to things like learning to ride or drive, showing, and advanced training of a horse, however, I've only told you what you need to know to be able to select and employ a professional. Proper selection of tack is most important and necessary for your safety and enjoyment, and that of your horse, but although it is not my area of expertise I cannot overemphasize its importance.

There are three things I believe you should know about tack. One is the general classification so you may be able to properly understand and talk about tack with the professional trainer and tack shop operator. A second factor, important to the health of your horse, is proper selection and fitting of tack so as to avoid injuring your horse or yourself. It is best to have a professional fit your tack when you are starting out.

The third is how to clean, care for, and store tack. Tack can be a vector of disease. Skin diseases can be carried from horse to horse by taking unclean tack from one horse to another. Respiratory disease can be spread if bits and bridles are switched from horse to horse.

Types of Tack

All tack is designed with two things in mind: use and appearance. For example, an English jumper saddle is made especially for that use, whereas the brass on draft horse harness is just for appearance. In addition, there are two general classifications of tack —riding and driving.

RIDING TACK

Riding tack is broken down into English and Western, with perhaps a third section, military or police tack, sometimes added to English. English has many subdivisions, such as flat, hunter, jumper, and racing, with numerous types of saddles and bridles and accompaniments for each. The classic Western saddle is a stock saddle, but there are special roping saddles, parade saddles, pony saddles, children's saddles, and so on. Western bridles are also different from English bridles, and there are variations of each for different purposes and different effects on different horses.

Military and police saddles are single-purpose saddles, usually with a high pommel front, but no horn, and often with an opening down the center, possibly better to ventilate the horse's back or to admit a high backbone on a thin horse. All military and police saddles have places for numerous pieces of equipment to be strapped or hooked. The classic McClellan saddle was used for artillery horses and could be combined with the harness of the horses pulling the caissons.

DRIVING TACK

Driving tack is divided into light harness and draft harness. From there on the types are innumerable. In the case of racing harness used on the Standardbred, there are so many special-purpose poles, straps, burrs, rolls, tie-downs, and hangers that the list would fill a whole book.

Parts of Tack

Starting with the front of the horse we'll list and give a brief description of appearance and/or use of various parts of tack.

HALTERS, ROPES, AND SHANKS

By tradition a newly purchased horse comes with a *halter* and a *leadrope*. A lead or tie rope is just that, a rope of hemp, cotton, nylon, or other synthetic material, and usually a snap to fasten it to the halter. It is meant both for leading and tying up a horse. A *leadshank* has a chain, one to two feet long, and a snap fastened to a strap of leather or nylon webbing. The chain can be snapped to the ring of the halter, or passed under the horse's jaw, through its mouth as a bit, under its upper lip, or over its nose, before being fastened to the D-ring on the right side of the halter. These positions are all meant to provide more control. If you are going to tie a horse use a lead rope, because tying with a lead shank can either break the shank or injure the horse. Lead shanks can be made for special uses. Polished leather ones are suitable for showing, while extremely long ones are used on stallions, allowing the horse to rear and not pull the shank out of the leadsman's hands.

Halters, also known as "head stalls" or head harness, are simply harness to go over the head, used to catch and hold a horse. They used to be made of leather or hemp webbing, and traditionally when you bought a new horse from a dealer it came with a cheap new webbing halter. Now many halters are nylon — extremely strong, but dangerous if a horse gets one caught, pulls back and slips off its feet, choking itself.

There are specialty halters such as heavy nylon webbing for stallions, tiny foal halters, halters with adjustable nose pieces to allow the mouth to be opened wider for dental work, beautiful leather show halters with matching lead shanks, and cotton and plastic rope halters that fit cows or horses.

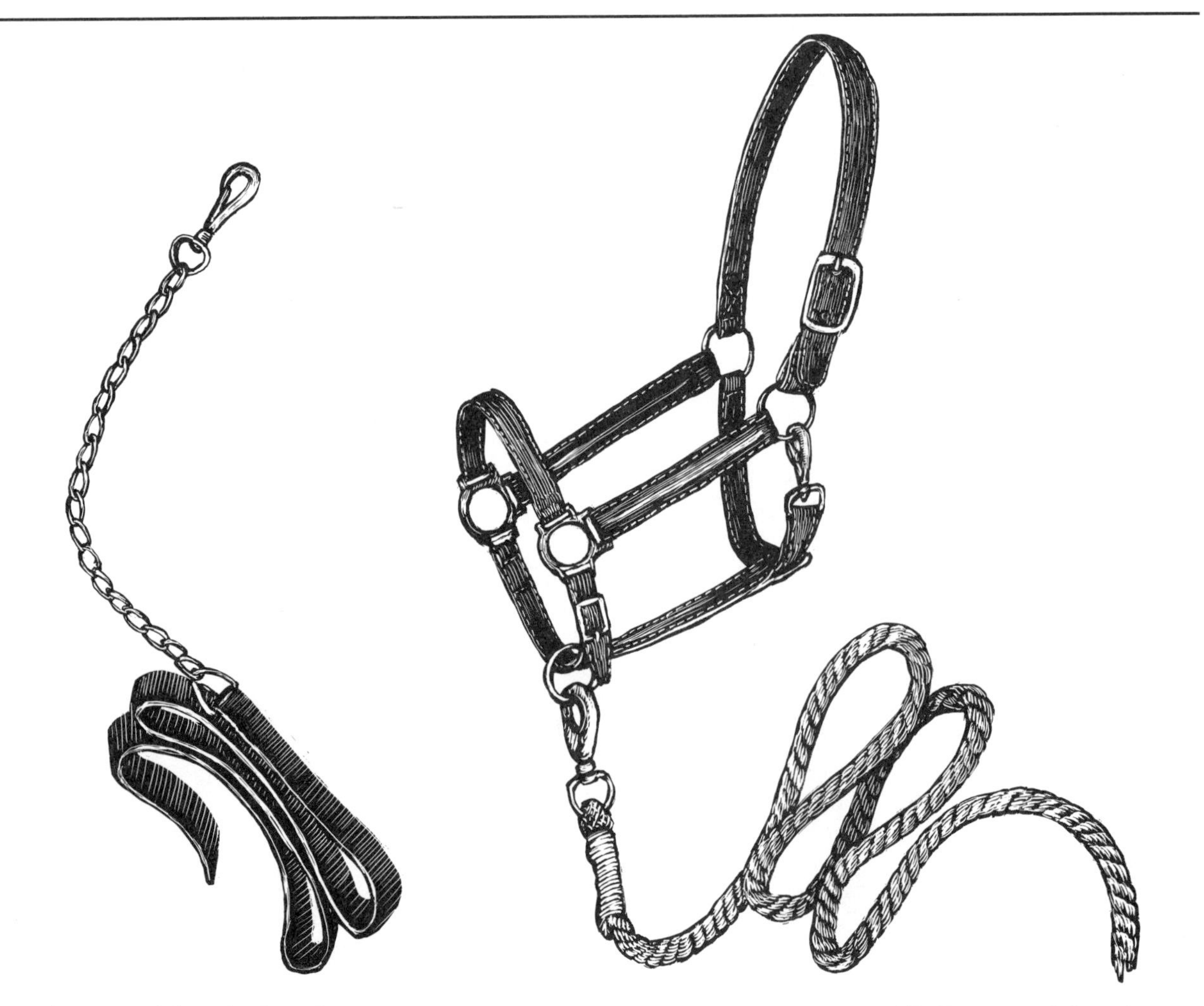

Halter and lead shank **Halter and lead rope**

BITS AND BRIDLES

Next on the head end are bits and bridles. For both driving and riding, English or Western, there are "open" bridles that leave the horse's face open. Each is enough different from another that they are obviously Western, English, or driving bridles. Bridles are made up of various parts (*see illustration*): *cheek piece*, *brow piece*, *head strap*, and *throat latch.* Particularly for driving horses, all sorts of things are put on the bridle such as *blinders* (and there are dozens of different kinds of blinders) to keep the horse from being frightened by things at its side, *shadow rolls* to keep it from seeing shadows on the ground (and jumping over them), and, of course, decorations such as are worn by parade horses.

English

BRIDLES

Western **Driving**

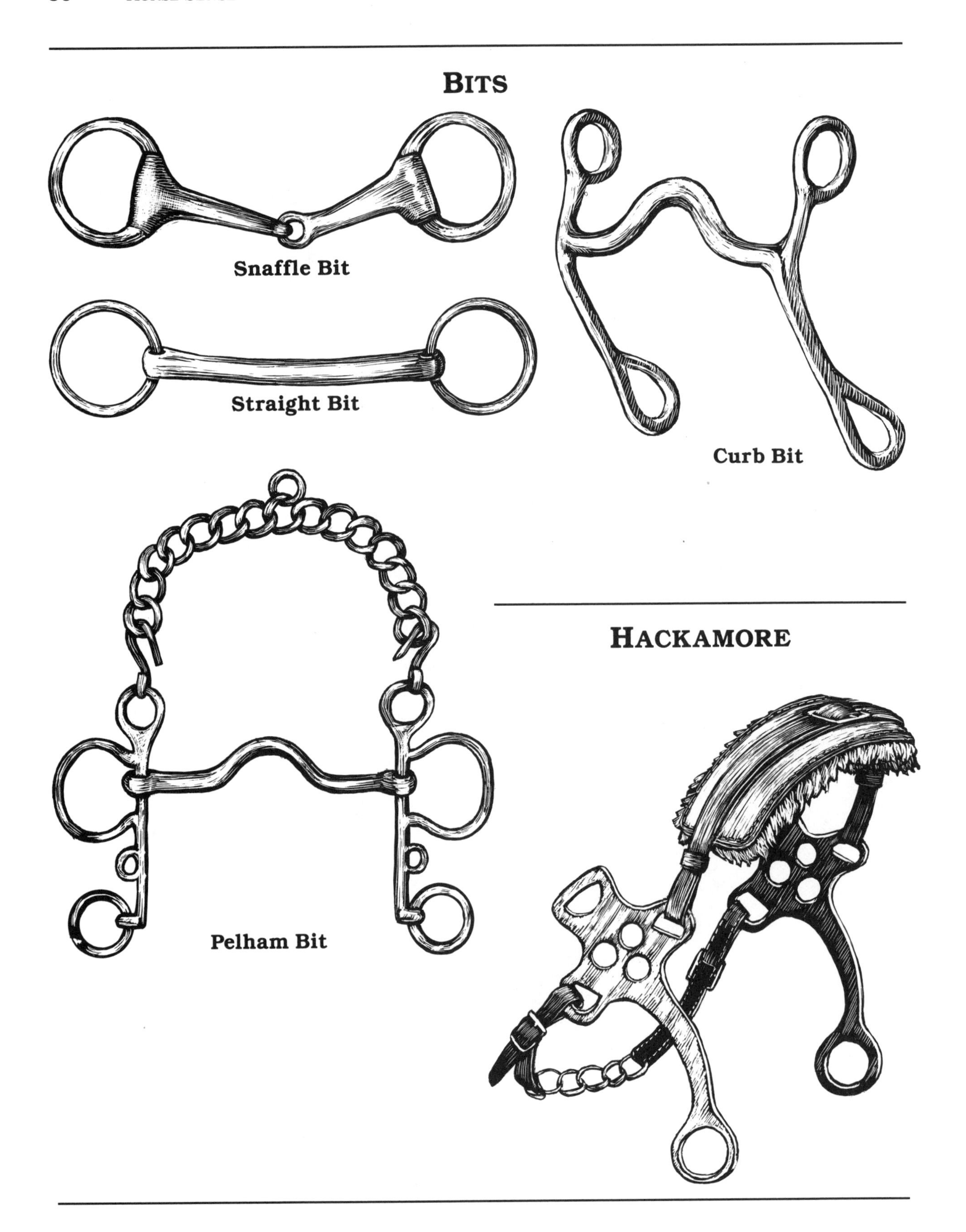
BITS
Snaffle Bit
Straight Bit
Curb Bit
Pelham Bit
HACKAMORE

Bits are so numerous that in old harness catalogs they ran page after page. Basic bits (*see illustration*) are *straight*, *snaffle*, *curb*, and a combination of the snaffle and curb called *Pelham*. They are usually metal, but can be coated with rubber. The most important thing about a bit is that it be as mild as possible while still controlling the horse, and that it not be too long or short. The difference between a 4¼″—too short, 4¾″—too long, and 4½″—just right doesn't seem like much, but can make a sore mouth, a horse pull to one side and carry its head turned, or the driver or rider lose control. It is essential that you have the bit properly fitted by a professional, preferably the horse's trainer.

Somewhere between a halter and a bridle, with no bit, is the so-called *hackamore* (*see illustration*). These are used on both Western and English riding horses with tender mouths, but rarely for driving.

REINS

Going back from the ring on the bit, which is fastened to the bridle at the lower end of the cheek piece, is the *rein* on a riding horse, or *line* on a driving horse. For generations, harness, including reins, has been made of leather, but recently some racing harness has been made of plastic, just as some race-horse reins are made of rubber. By tradition, reins on some Western bridles have been made of leather and are not fastened together back by the rider, where they are held. You will hear various reasons for

MARTINGALE

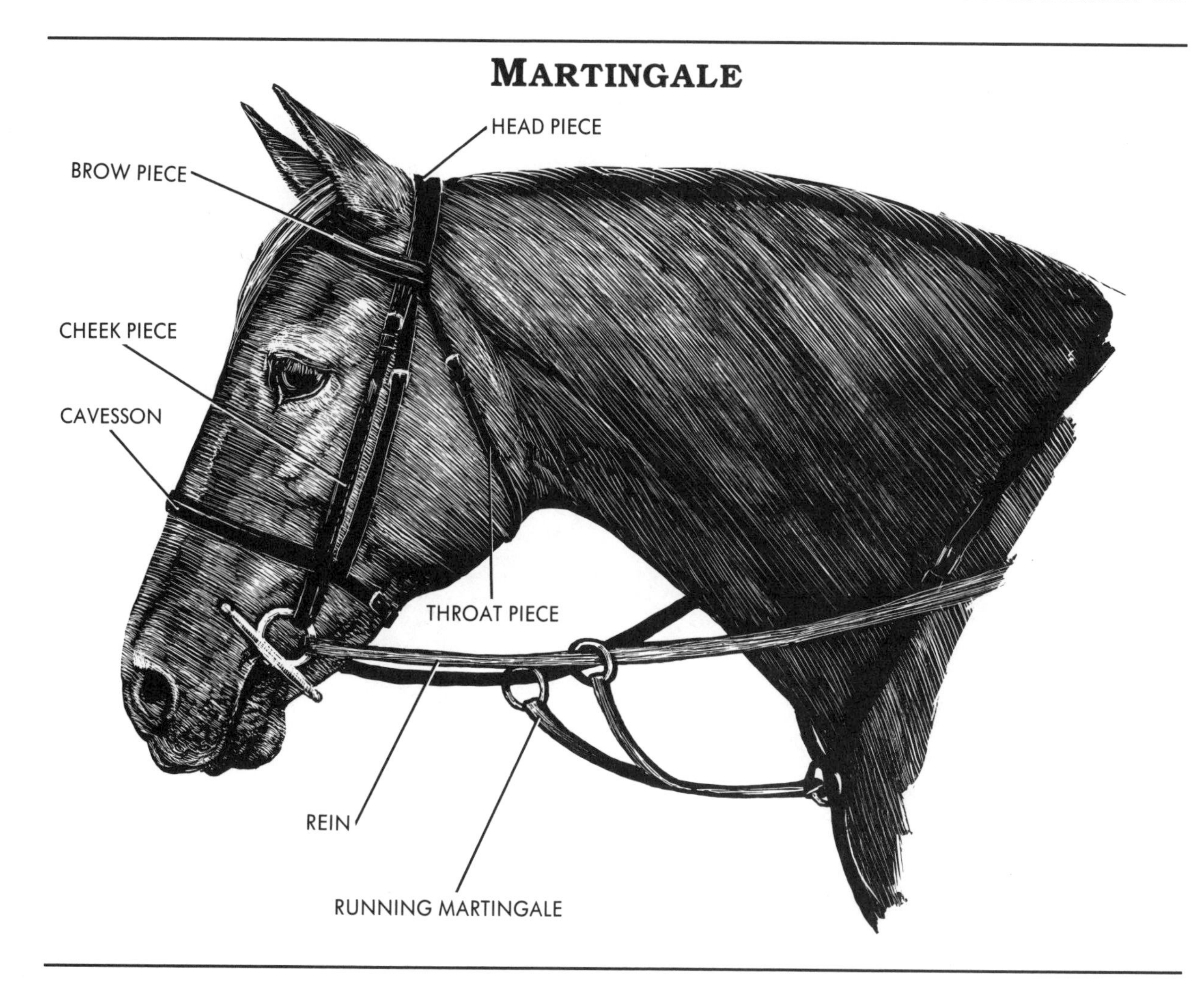

this, such as less risk of a steer's horns or a thorn bush catching on one rein, or less chance of an outlaw being able to hold the horse, or maybe it was just that rawhide only comes about so long. I don't know which is correct, but it makes for conversation when a Western rider and an English rider meet.

Parallel to the reins but shorter is the *check rein* or *check piece* which goes from the head piece back along the top of the neck to the withers. The farm horse cultivating corn wore a check rein to prevent it from eating the tops off the corn. The Standardbred pacer wears a check to hold its head up and keep it breathing better. A small woman or a child may need a check on their mount to keep it from reaching for a mouthful of grass and pitching them over its head.

Going back from under the horse's chin to the girth, and usually fastened to the top of the nose piece or cavesson (*see illustration*) is a strap called the *martingale*. This tends to keep the horse from throwing its head up. On the draft horse the martingale is fastened on top to the neck yoke and is used to help back up the load or hold it back going downhill. In this case it is fastened to two side straps at the area of the girth. A lighter version of the martingale is used on Standardbred race horse harness or light driving horse harness for horses that throw their heads.

LIGHT DRIVING HARNESS

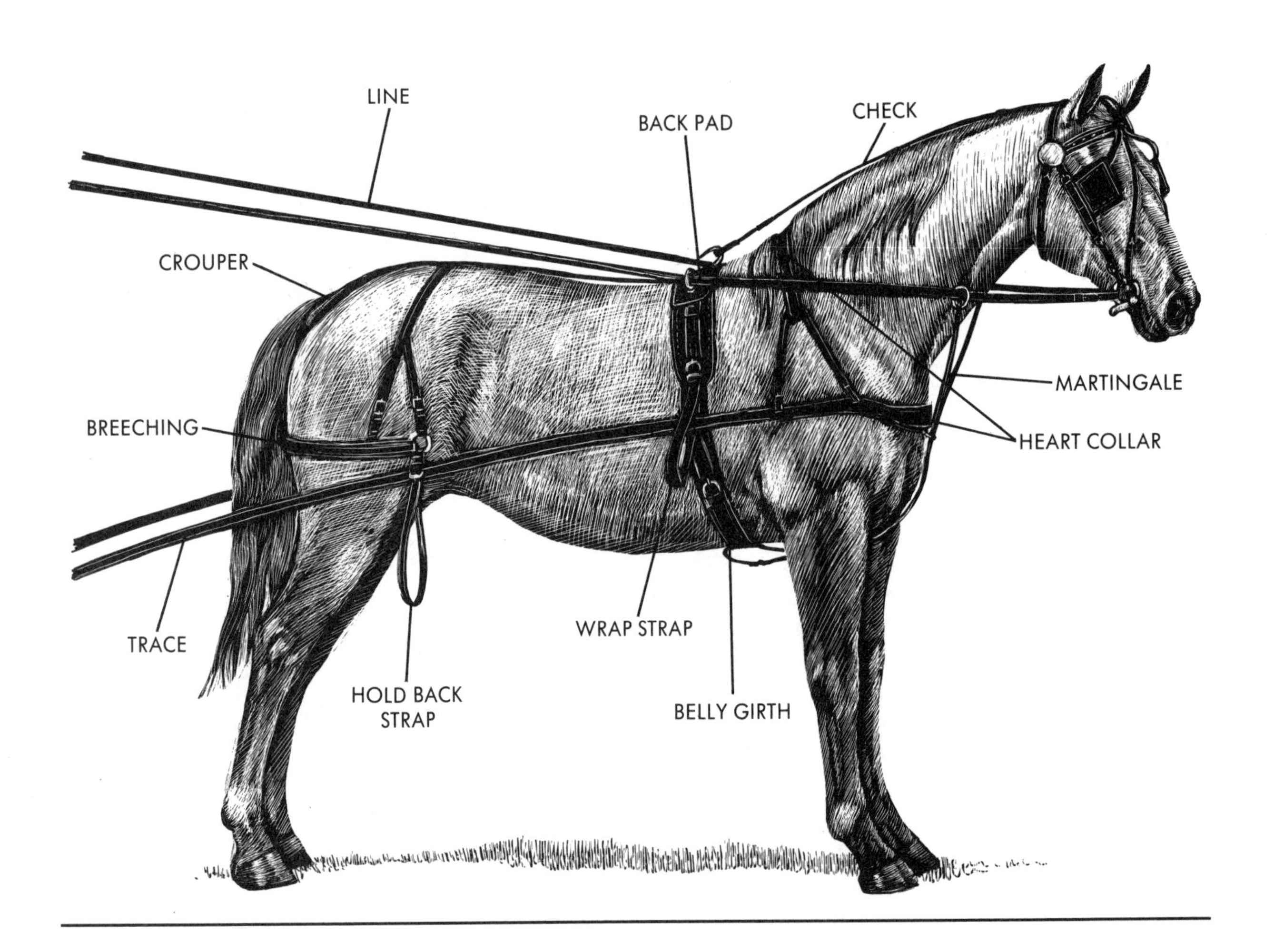

HARNESS

The next basic part and perhaps the most important part of the driving or draft harness is the *collar*, and there are two basic types. The breast or Dutch collar goes around the chest and directly to the trace on each side and then to the whiffletree. This type of collar is seen mainly on racing or light driving harness. It is held up by a strap going over the withers (*see illustration*). The work collar, or classic "horse collar," is used on heavy driving horses or draft horses. It usually fits over a collar pad and is sized by inside measurement from top to bottom, and sometimes from side to side.

Over and around the horse collar go the *hames*, matching wood and metal or all-metal pieces shaped to fit the groove in the collar. On the hames are fastened the traces. Some hames extend above the collar like horns and are decorated with a brass sleeve and knob. Others fit tightly over the top of the collar. In either case the top and bottom of the hames are held together with a *hame strap*. When removing the harness only the bottom hame strap is unfastened, while the top is left as is. The type of hames that extend up are often adjustable by moving the top hame strap up or down to a different set of holes.

To the hames is fastened the forward end

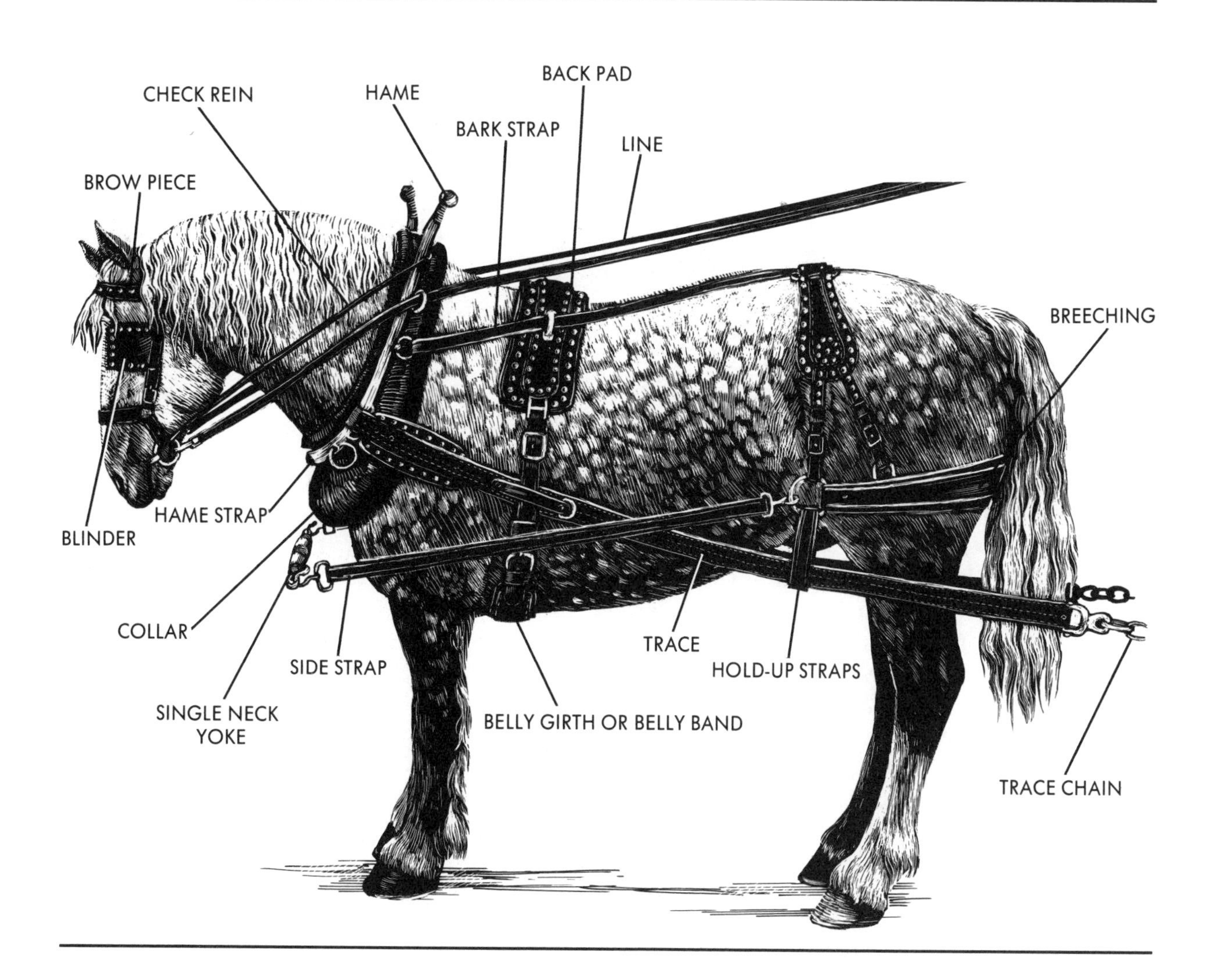

of the *trace*, the heaviest piece of leather in the draft harness. On draft harness the trace ends with a D-ring to which a chain with a hook is fastened. The chain of the trace is fastened to the *whiffletree* and can be adjusted by toggling the chain and hook. Light harness does not have a chain, but has holes in the traces to slip over the end of the light whiffletree.

Also from the hames hang either breast chains or straps to fasten to the *neck yoke*, which is used on a team pulling a wagon or anything with a tongue or pole. The neck yoke is a straight piece of wood, the size of a heavy ball bat, with a ring on each end for the breast chain or strap, and a large ring hung from the middle to go over the end of the pole.

From near the top of the hames goes a strap called a *back strap*, which goes over the top of the horse's back to hold up the *breeching*. The breeching is like a breast collar in reverse, which goes over and behind the buttocks of the horse to hold the load back or back it up by putting tension through the side straps down to the martingale and thus to the neck yoke.

In the single harness with a breast collar there is a full belly girth just behind the front legs. In the double harness there is often only a short girth from trace to trace under the horse behind the front legs.

The single harness has a breeching similar to the double harness, except it is usually fastened with a strap wrapped directly to the *shafts*. The shafts are parallel straight or slightly curved wooden or metal pieces that extend along the horse's side from the vehicle being pulled to just beyond the horse's shoulder. They are usually fastened on either side of the belly girth in the single harness.

Running along the back from the belly girth in the single harness is the *back strap*, which ends at the *crouper* (or *crupper*), a padded leather ring that goes under and around the tail. On occasion a saddle horse may wear a crouper with the back strap fastened to the saddle. This tends to keep the saddle from sliding forward.

As a point of interest, if you hang a horse collar in your den or tack room for atmosphere, hang it with the rear part away from the wall, and upside down. Any boy growing up with draft horses, when teamsters drove horses instead of eighteen-wheelers, learned that that was the only way to hang a collar so that it could dry properly and would not get out of shape. To this day, when I walk into a country inn and see a horse collar hung incorrectly I want to reach up and hang it the way I was taught.

SADDLES

Saddles, regardless of style, are fastened to the horse by means of a *belly girth*. Western saddles often have a wide webbing girth under the horse, just behind its front legs, with a ring on each end fastened to a girth strap (*see illustration*). English girths may be made of heavy canvas webbing or leather.

From both English and Western saddles hang *stirrup straps* of leather. There are nearly as many types of stirrups as bits, so we can mention here only that stirrups on the English saddle are known as "irons," and are made of metal. Often these are fastened to the saddle by a safety release. This device permits the stirrup strap to loosen and slide out if the rider falls off and catches a foot in the stirrup. Western stirrups are often covered from the front with leather hoods to prevent the foot from going through the stirrup.

Pack saddles, a variation of the Western saddle, have a breeching behind and a breast collar in front to stabilize them on the horse. They are still used in the west by camping and hunting parties.

SELECTION AND FIT

For me to try to tell you what harness and saddle are correct for your particular horse and situation would be just as foolish as telling you what type and size riding boots to wear. You can seek advice from a professional riding instructor, a driving-horse trainer, or a reliable tack-shop operator. As you gain experience you can use common sense to select items for your horse. The fit of tack and harness is just as important as the fit of your shoes. For example, a saddle too small for the horse or rider can cause saddle sores and girth sores that may never heal completely. The same can be said for collars; they must fit and be adjusted correctly.

OTHER USEFUL ITEMS

Boots. In addition to saddles, bridles, halters, and harness, tack includes items such as boots to protect your horse's knees, fetlocks, and coronary band. If a horse starts to damage its lower legs in jumping, boots

WESTERN SADDLE

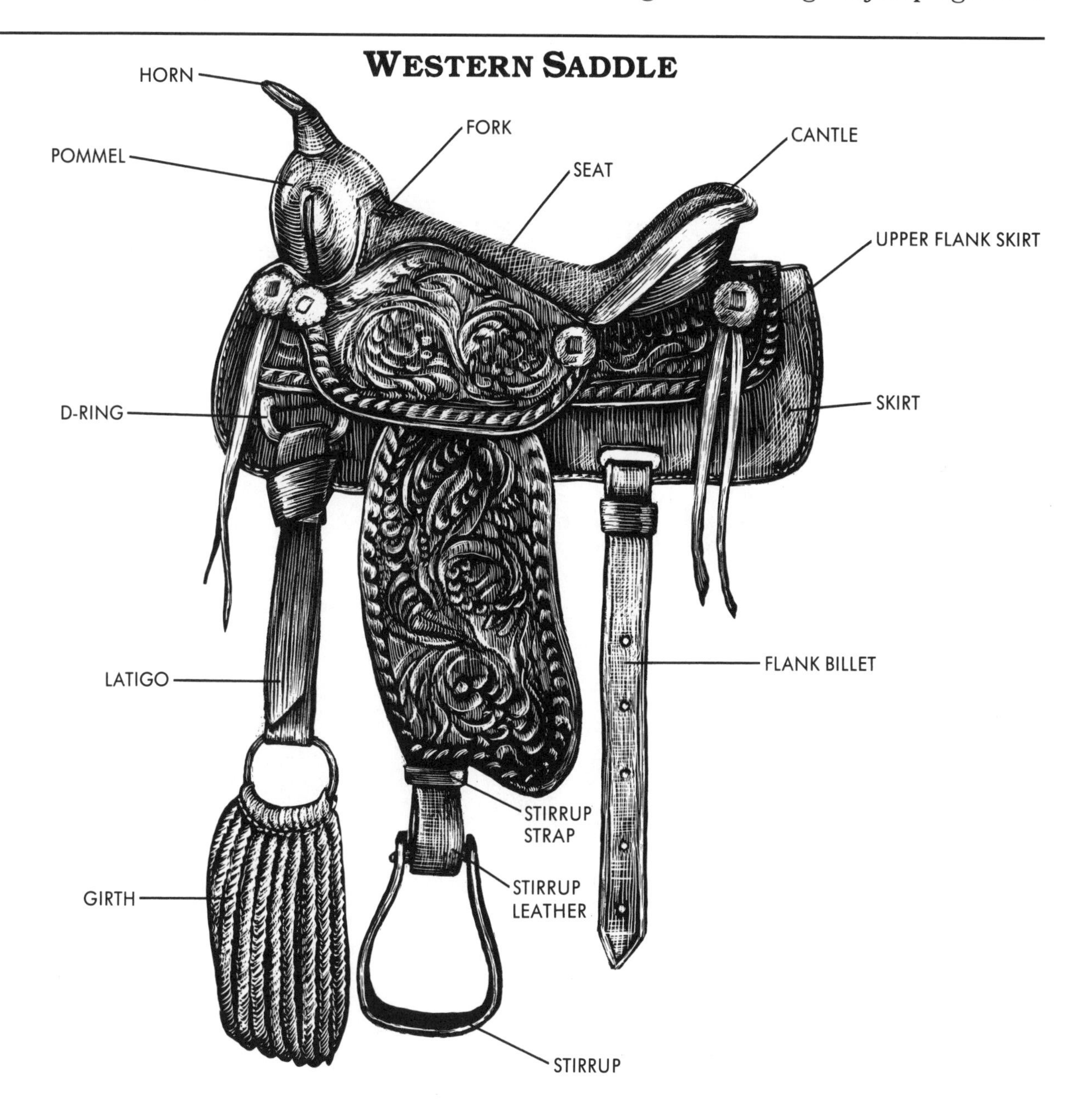

that cover that part of the leg may be indicated. Also, as mentioned in the chapter on shipping, rubber bell boots may be used to prevent a horse from stepping on its own heels and coronet when being transported.

Covers. *Sheets* are the lightest and most commonly used form of what most people would refer to as horse blankets. They are made of a light, soft, canvas-type material and may be put on a horse to keep its hair coat looking well while being shipped to a show, or to protect it from a cool breeze between show classes. *Blankets* are similar, but have a heavy, blanket-like inner lining for cooler weather. *Coolers* are really blankets made for a horse, and are put on after a bath, while cooling out during colder weather, or on a sweating horse between workouts or show classes. *Rugs* are the heaviest, and are used mainly on horses that have had their hair clipped short for winter training when they are turned outside.

If you bring a horse in cold and wet from outside in winter, or have a sick horse that is feverish and needs a blanket, have the blanket warming up on a *healthy* normal stablemate first before you put it on the cold or sick horse.

ENGLISH SADDLE

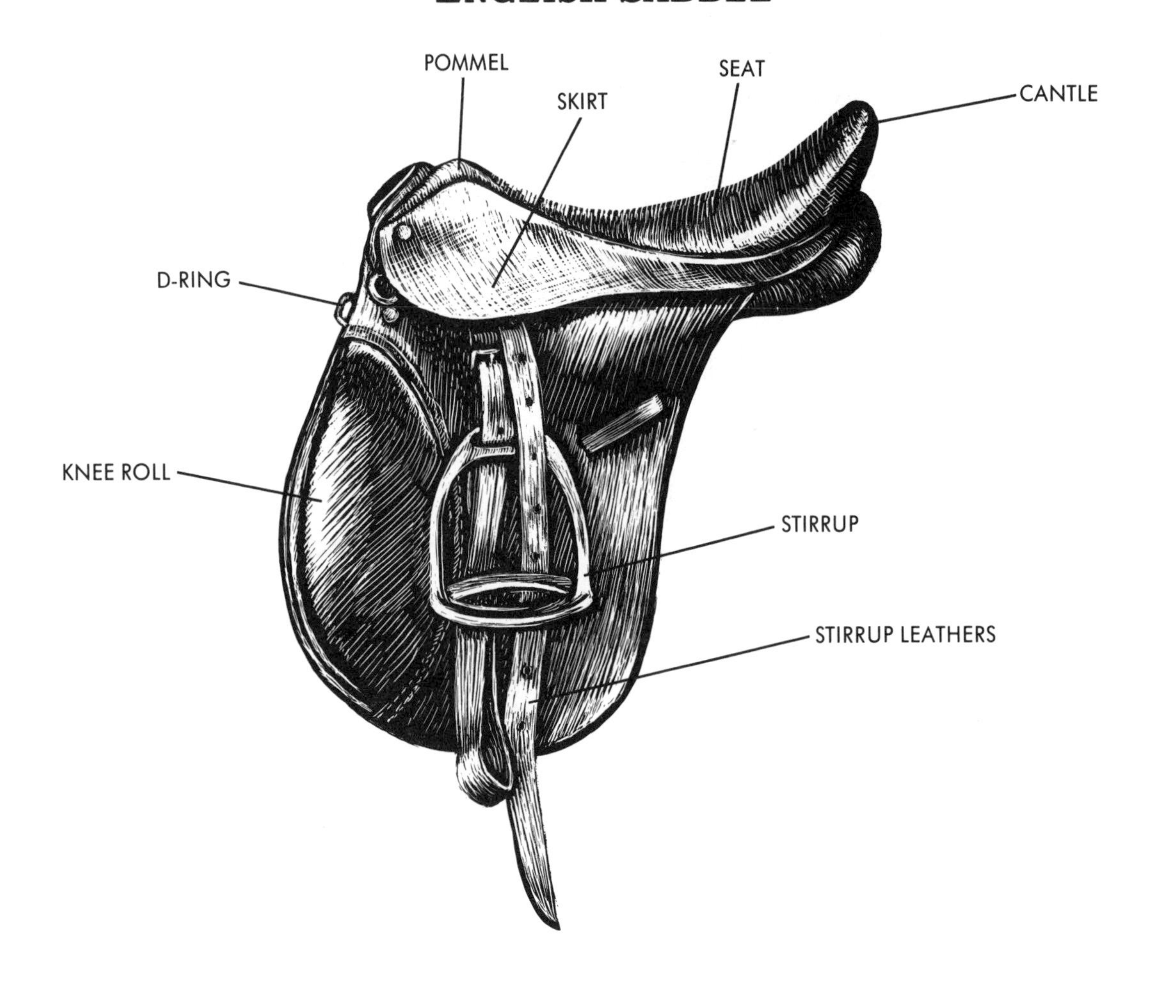

Care of Tack

When you remove tack or harness from a horse after a hot ride or workout it is just as important to clean and wash the tack as it is to wash the horse. Wipe the tack first with soft dry rags, then with damp rags. Every few days take the tack all apart and clean it with saddle soap, Lexol, or other product made for that purpose. If harness leather seems too dry there is nothing better than the liberal use of neat's foot oil. Soap and water are sufficient for plastic harness, although products that are used to protect plastic upholstery in automobiles should help lengthen the life of plastic harness.

When you remove a bridle from a horse after riding, wash the bit in clean water before putting it away so it won't be dirty when used again. Before bitting a horse in extremely cold weather, warm the bit in your hand or in warm (*not hot*) water before putting it in the horse's mouth.

Leather tack can deteriorate in storage faster than in use. Cold, dust, heat, dryness, and dampness are all enemies of leather. For long-term storage, clean and oil harness, saddle soap saddles, and store them in a dry dust-free place that is neither too hot nor too cold. Every six months or so stored tack should be taken out, dusted, and oiled. Wool blankets and coolers can be damaged by mice and moths if not put away clean and protected.

A HEALTH NOTE

We should know better, but we all make the mistake of switching tack from one animal to another without proper cleaning in the rush of a show, or meaning to do a favor for a friend by switching bridles or saddle pads. It is so easy to spread skin disease or even diseases such as strangles this way. You might lose a friend if you refuse to lend something, but if your friend's horse catches something from yours you'll lose him anyway. Regardless, don't be ashamed to wash and disinfect your tack when you get it back before you take a chance on bringing disease from some other animal to yours.

CHECKLIST FOR TACK

- **Have bridle, bit, and saddle properly fitted by a professional, preferably the horse's trainer.**
- **Clean bit and tack after you remove it from your horse.**
- **In cold weather, warm the bit before placing it in the horse's mouth.**
- **Do not switch tack from one animal to another without cleaning it properly in between.**
- **Store tack and other gear carefully.**

CHAPTER VIII

METTLER

FIRST AID

There was a time *when a veterinarian's telephone was answered either by the veterinarian personally, or by a family member or employee who could give the caller advice on what to do until the veterinarian arrived. With the use of electronic devices for communications this person-to-person relationship may not be as evident as when veterinarians used to be on call twenty-four hours a day, seven days a week, not even taking a day off. Yet it is still important.*

When electronic telephone-answering devices first became available, many veterinarians found them a reliable way to assure that their telephones were answered when no one was in the office. In an equine practice of only one veterinarian, after all, it was more practical for an employee to accompany the veterinarian on calls rather than remain in the office to answer the phone. With this help the veterinarian could not only accomplish more but also keep the better health records that are required on each horse for preventive care. Since most stables have a telephone, the veterinarian or his or her assistant needed only to call in and check the messages on the recorder every half hour or so.

Some clients, of course, detest answering machines and refuse to talk on them, although most get used to them. Some, like Edwina, never got the hang of using them, and never seemed to remember to call in the early morning or any other time when I was able to answer the phone.

My outgoing message was: "This is Dr. Mettler's office. I am unable to answer the phone at this moment but will be checking the recorder regularly, so at the sound of the tone, please tell us who called, and leave your phone number if you wish to be called back, and/or a message. Thank you for calling. Here is the tone."

When I would call back, if there was a message such as, "This is Tom Weaver; Satin Lady has colic," I'd put off other calls and go there immediately. Or the message might be, "This is Bob O'Brien; I just sold a horse to go out of state, will you stop when you're by here and Coggins him." If near Bob's I'd stop in, or else take care of it that afternoon or the next morning. Another message might be, "Sue Burke, 351-2343; please phone me about setting up a time to geld a new colt I bought." This I would put off until I returned to the office.

No matter how many times I explained this to Edwina she would always leave the same message — simply "Edwina Edison, 351-3341" — very matter of fact, with no indication whether it was urgent or if she just wanted to visit about the prospect of buying a new horse.

One spring morning I was running way behind on routine calls and had missed breakfast because of two colics at opposite ends of my practice area. At about 10 A.M. there was the usual message on the recorder — "Edwina Edison, 351-3341." At 3 P.M. I stopped at the local diner for coffee, figuring it would suffice for missing lunch and hold me and my assistant until dinner. While my assistant ordered the coffee I went to the public phone and checked the recorder. I returned a couple of calls that sounded urgent, and then, instead of waiting until evening, I phoned Ms. Edison.

Her small daughter answered. "Mommy is out in the stable holding something on Rhubarb so he won't bleed any more, so she can't come to the phone. I'll tell her you called."

I replied, "Tell her I'll be right there," but the child had already hung up. The coffee was paid for and left on the counter while my assistant and I ran for the practice truck.

When we arrived at the Edison farm there was Edwina, appearing calm and cool as ever, holding a sanitary napkin against the gaskin of her child's pony. In an unexcited tone she explained that McIntosh, the big Thoroughbred gelding, had just been shod with caulks and had kicked Rhubarb, opening a deep gash in the fleshy part of the gaskin. At first she thought the wound would stop bleeding, but after she had phoned it kept getting worse, so with no first-aid supplies on hand she had tried using the sanitary napkin, holding it on since she had no bandage that would keep it there. This worked fine, but every time she took the pressure off it started to spurt again. She sent her daughter, Melissa, to the house for more pads and had just put on the last one when we arrived.

A little local and four stitches later, the

bleeding was stopped, and after a toxoid booster and another lecture on how to leave a message on the recorder we were on our way.

Three days later on an equally busy morning a message on the recorder said, "Edwina Edison, 351-3341." Thinking that perhaps Rhubarb had developed wound septicemia, I skipped the next call and drove directly to the Edisons' place. No one was home, and we found Rhubarb in his stall looking his usual healthy Shetland self.

That evening I called Edwina and she had to think a moment to remember why she had called. Then she said, "Oh yes, I wondered if you had any idea where I might find a child's saddle for Rhubarb?"

Now, better than devices just to answer phones and record messages, machines combining radio and telephones assure that in a true emergency the veterinarian can be reached immediately or the client can be put in touch with someone else for veterinary help or advice. With the use of these new communication systems today's veterinarian, better educated and better equipped than his or her counterpart forty years ago, is able to spend more time on preventive veterinary medicine.

To make preventive veterinary medicine work there has to be a strong personal relationship between the veterinarian and the client. The client cannot wait until there is an emergency to start looking for a veterinarian.

Today's better-educated veterinarian expects a better-educated client, one who takes care of horses so that most emergencies are avoided, one who knows an emergency from a routine call, how to care for minor ailments and injuries, how to give first aid until the veterinarian arrives, and how to follow the veterinarian's directions for good nursing after care. Again comes the need for the personal relationship. The veterinarian you choose must take the time to listen to you and then instruct you on care of injured or sick animals and, most important, discuss measures that might be taken to prevent a recurrence.

Selecting a Veterinarian

Injuries often occur when horses are being moved to a new place or during their first few hours there. Before you move a new horse to your stable you should have made some contact with a veterinarian. As mentioned in Chapter I, having a veterinarian check a horse prior to purchase is a good idea, not only to learn more about the horse, but also for you and the veterinarian to meet each other. Then when you call you won't be just another "shopper," but a known client with a horse and a problem needing attention.

How do you make your choice of a veterinarian? Ask other horse owners in your area for suggestions, but make your decision based on your own personal preference. You certainly want someone who is both experienced and informed on the latest in equine medicine. The veterinarian who works at the local racetrack or one who cares for the horses at a local breeding farm may be best for that type of practice, but do they consider the care of backyard horses important? Are they willing to take the time to explain to you all the little things about horse care that their professional horse clients have known for years? Would a younger veterinarian who combines small-animal and equine practice be better suited to your needs? Or would an older practitioner who has worked with horses for years in a general large-animal practice be your best choice? The choice of a veterinarian, a farrier, a feed supplier, and a riding instructor is more a matter of personality, yours and theirs, than any of us care to admit.

Once you have made that choice you don't have to stick with it permanently, but before making a change be sure you can get service the next time you have an emergency. Clients who call a different veterinarian every time they need one, or use a status veterinarian from miles away for routine work and call the local veterinarian only for emergencies, sooner or later find themselves some summer Sunday afternoon with an emergency and no veterinarian.

Coping with Injury

BLEEDING

As in first aid for humans, the first thing one must do for an injured horse after calming it and securing it where it is not afraid or able to hurt itself again, is to stop the bleeding. Bleeding is serious when arterial bleeding is evident by spurting or when deep wounds flow a steady amount of blood. Unless spurting is obviously from an artery 1/16 inch or more in diameter don't do anything. Most bleeding will stop by itself in five minutes if the horse is kept quiet, and bleeding will wash dirt and bacteria from a wound.

Then use direct pressure, usually not with a tourniquet, but by holding a sterile gauze pad where the blood seems to be coming from, or by packing a deep wound with gauze pads. Don't apply water, hot or cold, or strong astringent powder or, worse yet, cobwebs, flour, or wound powder containing talcum.

THE STABLE MEDICINE CHEST

Ask your veterinarian what you should have on hand for emergency and routine first aid. Most of these supplies can be purchased at the veterinary clinic. A well-stocked stable medicine chest might contain all or some of the following. These are brands I am familiar with and recommend, but there may be others equally good.

Essentials:

- Nitrofurazone aerosol wound spray (Topazone or equivalent)
- 2-4-ounce bottle tincture of iodine
- One package sterile gauze pads
- Small plastic pail (8-quart)
- Roll of non-sterile cotton
- 2-4 3″ rolls of Vetwrap or Elastikon
- 5″ fever thermometer
- One pint to a quart disinfectant (Betadine, Nolvosan, or Lysol)
- One or more cakes Ivory soap
- Clean paper towels

Handy, but not essential:

- Four leg wraps
- Four quilted pads in clean plastic bag, with safety pins
- Small bandage scissors
- Roll of plastic tape
- Oral colic mixture or pint of Maalox or other antacid
- Plastic or rubber and aluminum dose syringe
- Nitrofurazone or other wound ointment
- Chain lead shank with nylon strap
- Aluminum "humane" twitch (clamp-style)
- Betadine solution
- Supply of sheet cotton
- Other things such as "bute," aspirin, etc., as suggested by your veterinarian

---

LEG WOUNDS

For wounds at the hock or knee, or below either, try wrapping, using only gauze, cotton, and pads, applied with plenty of pressure. For wounds at the bulb of the heel or back of fetlock that are too low to wrap, try bandaging the entire hoof. Any wound serious enough to require this much attention, or any puncture wound more than skin deep, requires veterinary attention, if for no other reason than to give a tetanus booster.

Any horse with a puncture wound, or deep wound, that has not had a tetanus toxoid injection within three months should have a booster. All horses should have tetanus immunization given by a veterinarian by three months of age, followed by a yearly booster.

Most scrapes, abrasions and shallow cuts need only a little wound spray to reduce infection and don't need to be seen by a veterinarian. If a small wound or an abrasion is full of dirt don't be afraid to scrub it clean using sterilized gauze pads and a Nolvosan or Betadine solution.

Wounds to the knee or hock or over other joints may appear minor but can be very serious. A wound over a joint such as the knee, which drips clear or straw-colored fluid, indicates an open joint. These should receive veterinary attention immediately. Your suspicion of an open joint should be made clear to the veterinarian or his or her answering service.

Other wounds over the knee or hock should be bandaged if at all possible while

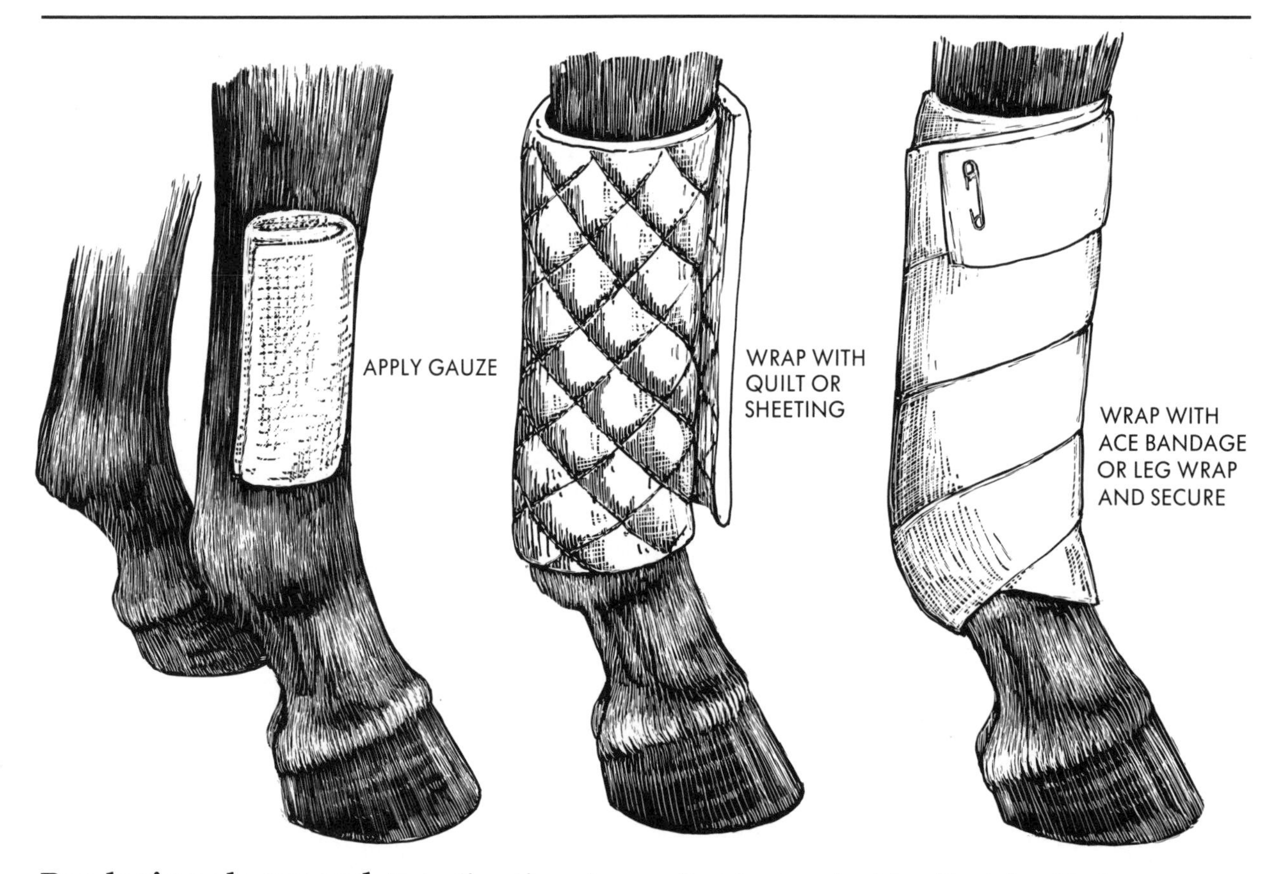

Bandaging a leg wound. **Note direction of wrap. Bandage on inside of leg should always wrap in forward direction so that if horse bumps it with opposite hoof as he walks it will tighten, not loosen, bandage.**

awaiting the veterinarian, using furazone, sulfa, or other antibacterial ointments under sterile gauze, covered with cotton and then secured by Vetwrap or Elastikon. Such wounds should have bandages changed every day or every other day. If they appear to be slow healing or causing lameness call your veterinarian.

To wrap a leg, cover the wound with sterile gauze, then wrap with sheet cotton, cotton pads, or even baby crib pads, with the wrap in the forward direction on the inside of the leg. Follow the pad with Ace bandage, Elastikon leg wrap, or Vetwrap, and secure if necessary with safety pins, Ace bandage clips, plastic tape, or Elastikon. On knees or hocks, figure-eight the bandage, and if necessary apply a second supporting or "standing" bandage to the cannon area below to keep the hock or knee bandage from sliding down the leg. After applying a bandage to the front legs watch the horse to keep it from chewing the bandage. To prevent this one may cross-tie the horse or apply a bib or cradle.

EYE WOUNDS

Wounds around the eye, particularly torn eyelids, caused by horses rubbing their faces against objects and catching an eyelid, are common and serious. Still, no matter how serious they appear, it is amazing what a veterinarian with a little patience and your help can do to repair these if called early. For first aid don't do anything but call the veterinarian. However, determine where the horse injured itself and remove the hook, wire, nail, or whatever caused the injury. Torn wire mesh over stall windows is a common source of such injuries.

LARGE-MUSCLE WOUNDS

Deep wounds to the buttocks of horses from kicks, and to the pectoral area between the front legs from straddling fencing, are common. Other than giving a tetanus immunization, your veterinarian may decide to heal them "Colorado"-style by just letting them fill in naturally, with no further treatment except washing out with Betadine or Nolvosan and water, or even hosing clean with tap water and using a little wound spray. This may cause the veterinarian to appear lazy, but in the end healing may be more scar-free. If a scar or granuloma (proud flesh) does result, surgery can always be done to correct it, whereas if the wound is sutured shut it may develop an abscess or restrictive scar tissue. To paraphrase an old saying, there is more than one way to heal a horse's wound.

HOOF WOUNDS

As mentioned in Chapter V, puncture wounds to the hoof sole should have tincture of iodine applied at once and, in most cases, require a tetanus booster given by your veterinarian. If the wound is visible the veterinarian will usually make the hole big enough to drain well. Here, as with large-muscle wounds, different veterinarians have different views as to whether such feet should be bandaged. The way your own veterinarian chooses to treat the wound is the best way. Wounds to the coronary band may appear minor but are often serious. Clip away the hair and loose skin or flesh with sharp cuticle scissors and wash the wound with Betadine. Leave it open and keep it clean, or call your veterinarian for advice.

If a horse has an undiagnosed lameness for several days or more and then suddenly develops what appears to be a wound at the coronary band and becomes *less* lame, you and your veterinarian did *not* miss the wound. This is what horsemen refer to as "popping a gravel." The horse had an invisible wound to the sole of the foot in the area of the "white line" (see diagram in Chapter VI). This wound formed an abscess that traveled *up* the wall of the hoof and opened at the coronary band. It is not too late to give a tetanus booster at this time.

BRUISES

Serious bruises to the legs and joints of horses, with little or no broken skin, are common. Cold water shower is the first aid of choice, followed by dry wraps. Your veterinarian may advise hot wraps or sweats used alternately with the cold later, but you can never go wrong using the cold until told to use hot. Simply cross-tie the horse, or have someone hold it if it is nervous, and let cold water from a hose run down over the affected area for a quarter to a half hour. After the first time you will be able to hold the horse with one hand and the hose with the other. If your veterinarian suggests a "sweat" on a horse's leg he or she means to coat the dry leg with nitrofurazone ointment (or some other product of choice) and wrap the coated leg with plastic such as Saran wrap. Then wrap the plastic with a layer of cotton, or cotton pad, and secure with an elastic-style bandage or leg wrap. These are usually changed every twenty-four hours.

DMSO and Counter-Irritants. Liniments, ointments, and other mild counter-irritants are often used on horses. Do not use them unless you are doing so on the advice of a veterinarian. Never use anything such as DMSO or any other product on a horse before a veterinarian is called and asked about it. Recently several horses died when DMSO was applied after a counter-irritant containing mercury had been used a day earlier. The DMSO caused the mercury to be absorbed, which poisoned the horses. "Old horsemen" have lots of remedies that work for various things, but if it is a secret formula, forget it — such things went out with other folk medicine.

COLIC

First aid for colic and other systemic diseases or problems will be discussed in a chapter dealing with those conditions. In any situation, if you are in doubt, phone your veterinarian with an accurate description *as you see it*, and find out what he or she thinks.

Other Reasons to Call a Vet

Other than lameness or obvious injury, your reason for alarm might be that your horse is:

- refusing to eat
- eating only hay
- eating only grain
- slobbering when eating
- coughing
- spitting out wads of hay while chewing
- showing change of attitude, such as depression, hanging head, drooping ears, nervousness
- discharging from nose and eyes
- showing a change from normal appearance of eyes
- sweating for no apparent reason
- acting uneasy
- getting up and down
- kicking at belly
- straining and/or squealing
- having diarrhea or change of consistency of feces
- not passing any feces
- urinating frequently or appearing to attempt to urinate
- feverish (see next section)

You may notice that I have not mentioned color of urine, because horse urine can vary so much in normal appearance that unless associated with other obvious disease one does not normally consider just a change in its appearance as cause for alarm.

Your Horse's Vital Signs

Temperature. In addition to the above signs, you can easily take the horse's temperature with a rectal thermometer. Simply take hold of the horse's tail and insert a lubricated five-inch thermometer three inches into the rectum. Be sure, of course, that the thermometer is shaken down below 96° before inserting it. Lubrication may be Vaseline, Ivory soap, or, if it will not offend anyone, simply spitting on the bulb end of the thermometer. Unless just worked or standing in hot sun or a hot, airless barn, horses' temperature should be at or below 100° F (37.8 C). A temperature of 102° or above on a resting horse is cause for alarm but, unlike humans, normal temperature varies so much in horses that a diagnosis on temperature alone is not attempted.

Pulse. A horse's pulse may be taken by listening just behind its elbow or, with practice, by feeling under the jaw as shown in the diagram. Normal pulse should be 30 to 40 beats per minute but, like temperature, can vary tremendously under normal conditions. Over 60 on the resting horse is cause for alarm, 80 to 100 indicative of serious trouble, and over 100 can mean a fatal colic.

Respiration. Respiration can be counted from a distance and should be judged not only by its frequency but also by its character. So-called normal number of respirations per minute in the horse is twelve, the same as in the human, and often this is difficult to detect. When respiration is rapid, labored, or noticeably different than usual for no apparent cause, it should alert a horse owner to trouble. A higher than normal rectal temperature, along with abnormal breathing or any other signs such as refusal to eat, and running nose or eyes, are certainly reasons for alarm.

Never call your veterinarian and leave a message such as "I have an emergency," with no explanation, or ask for a call back with no indication whether it is a life-or-death matter or just a desire to set up a time for worming. Find out from your veterinarian if there is a particular time when he or she prefers to receive calls, and when is a good time to discuss things with you. It can save both of you a lot of difficulty and preserve the personal client-veterinarian relationship that is so crucial to success in caring for your horse.

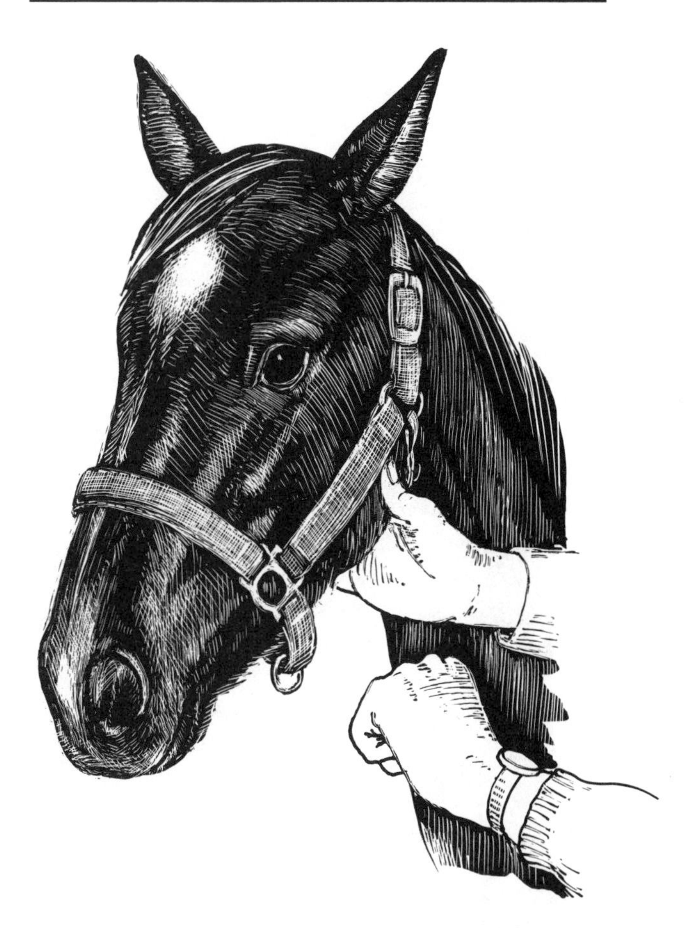

Taking a horse's pulse

CHAPTER IX

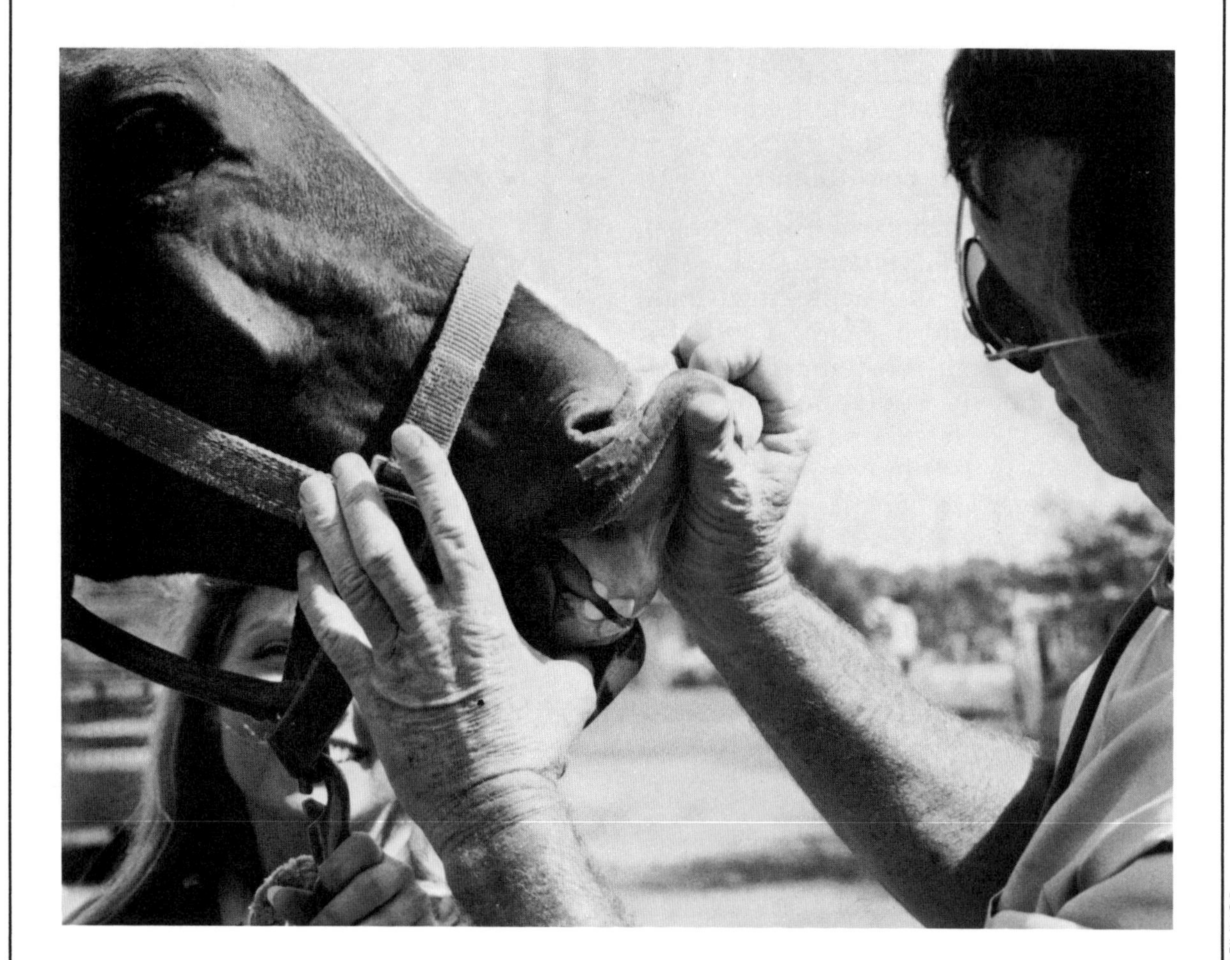

BONNIE LINDSEY/HORSEMAN MAGAZINE

SHOTS, TEETH, WORMING, AND COGGINS

Mr. ***Tag Along*** *was a riding-camp reject. All that saved him from being sent to slaughter was the sixth sense, or perhaps we should say horse sense, of a dealer, and the dreams of a thirteen-year-old girl.*

When I first saw him he was in a straight stall in a dealer's barn. He was so thin that his hip bones seemed about to protrude through spots of hairless skin. The space between the large hamstring muscles, sometimes called the poverty line, was so wide and deep you could hide your hand in it. He was a red chestnut, dull of coat, with one white sock and a white star and stripe. The stripe was hairless over his slightly Roman nose. His soft and intelligent eyes were the only thing that told of the real horse inside this sad-looking frame. He turned his head and looked back at Jeannie, small for her thirteen years, with love and expectancy.

My daughter Jeannie had come to the dealer's stable with me to find a suitable horse for continuing her 4-H experience. On that Sunday after Labor Day there were several decent-looking horses in stalls and some outside in paddocks, plenty to choose from, but Jeannie had eyes only for this sorrowful-looking scarecrow of a horse, and his eyes drew her like a magnet.

Before I could stop her, she spoke to him, walked into his stall, and was stroking his neck and talking to him as if she'd met an old friend. A moment later she was picking up his left front foot. In fact, as I recall, she just ran her hand down his leg and said, "Let me see your foot, boy," and he picked it up. "He's barefoot, Dad," she said. "Good feet, but need trimming."

Jeannie untied the horse and backed him out of the stall, handing me the tie rope. I looked into his mouth and, to my surprise, guessed he was no more than thirteen years of age with good teeth that appeared to have just been floated. Then Jeannie picked up the near hind leg, went back to the off front, and was just checking the off rear when the dealer walked in.

He took in the scene and, as though it were a line in a play, said, "That isn't the horse I had in mind for you, but I believe he's the one you're looking for. Would you like to try him?"

The dealer saddled the horse. Jeannie mounted and rode him slowly around the outdoor ring, and then went into a trot and slow canter. "Will he jump?" she asked.

The dealer hesitated a minute, and then said, "I've seen him jump, but I'll be honest with you. The kids at the camp couldn't even ride him. When I picked up the rest of the horses last week they told me he was a renegade. All he wanted to do was rush back to the barn, and he dumped every kid who tried to ride him.

"Last spring, though," he went on, "when I bought him in the city, I saw a woman no bigger than you jumping him. He looked great but was a little thin, maybe due to worms. I told them at the camp to have their vet worm him and check his teeth. They did neither. They told me their shoer couldn't even pick up his feet, he was so mean. They didn't want to spend any more on him and said I should take him right to the killer. But a horse with eyes like that...I couldn't do it.

"Now I come in here and see you, all of seventy-five pounds, picking up his feet! If you can pick up his feet you can jump him, but wait until you have him home and have a proper instructor."

The dealer removed the saddle and Jeannie started back into the barn, saying, as I'd advised her, that she wanted to look at some other horses before deciding. After looking at and trying several other far better-looking horses and checking on their prices, which were all over $500, top price for that year and that season, she asked to look at the chestnut again. This time she took him out of the stall and just led him around. Noting the way he followed her, she unsnapped the lead rope and he never changed. She turned, the horse turned; she stopped, the horse stopped, as though she had him on an invisible lead.

"Look dad," she called. "He's Mr. Tag Along." Then she added, "I want to take him home."

The dealer looked at us for a moment,

then said, "That horse could be worth more than all the rest put together, but I know if you don't take him I won't be able to give him away, and I'll be stuck with him all winter. You take him home for $175, and if he goes sour for you like he did for the camp bring him back and I'll give you your money back."

Thus, Mr. Tag Along came to be. When we took him home we were greeted with ridicule. "What a specimen for a veterinarian's daughter to own!" We wormed him and rechecked his teeth, which needed no further treatment. I wish I could tell you that, like Velvet, he blossomed into a beauty in a few short weeks. In those days wormers were not as efficient as they are today and we had to worm him every thirty days until spring and feed him enough grain and high-quality hay to keep a race horse going. Still, it took six months before the "poverty line" filled in on his hind quarters. It took another year before Mr. Tag Along took on the bloom that I have always taken for granted in horses that I've cared for.

By the time Jeannie was a skilled equestrienne in her midteens, Mr. Tag was technically an old horse and should not have had the stamina to compete and win in competitive trail rides and show-jumping. Still, he did, and as each year went by he seemed to fill out more, and until "retired" at age 18 never looked half that age. During Jeannie's college years she rode him occasionally for old times' sake, generally going out alone on him when she was home. I believe she and Tag could communicate as two old friends who just enjoyed each other's company.

When Tag and Jeannie were each twenty-eight, and she was practicing law 100 miles away, Tag was the only horse left at our place. For the first time in his life he actually appeared fat. One fall day, while fixing a board on the fence near the open shed where he spent most of his time, I realized he had not finished his morning grain. Checking him over, I could find very little wrong at first except the pale color of his mucous membranes and a slightly fast pulse. When I took his temperature the thermometer came out covered with tar-like blood and read below normal temperature. Apparently Tag was bleeding from an ulcer or eroded tumor in his digestive tract, a not uncommon cause of death in old horses.

Tag didn't appear to be suffering, and showed no colicky signs. Although he wouldn't eat hay or grain he took an apple from my hand and chewed it slowly.

A phone call to Jeannie confirmed that she could be home in two hours. Meanwhile, a friend who operates a backhoe came and dug a hole near the spot where Velvet and Rusty, our other two horses, were buried. When Jeannie arrived it was nearly dark, but Tag was still on his feet, standing in the same spot under the shed where he'd been all day. She went into the shed and I left them alone for a few minutes until she called, "Okay Dad."

She walked out of the shed, through the open gate toward the gaping hole, with Mr. Tag following her as of old. Jeannie rubbed his nose while I administered a concentrated dose of barbituate. There was no sound, no groan. Tag simply lay down. We covered him with one of the blankets he and Jeannie had won at a show. The next morning he was nudged into the hole by the backhoe and covered.

The barn and shed stood empty for many years, but when I come home after dark on a fall evening, as I get out of the car to open the garage door, I still listen and imagine I hear that familiar nicker from under the shed. But there is only silence, and my happy memory of Mr. Tag Along.

Problems with horses, as with most anything else, are far easier to prevent than to correct once they happen. With cattle, poultry, sheep, and hogs, good livestock practice mandates a herd or flock health program, sort of like preventive maintenance on your automobile. On horse breeding farms, at racehorse stables, and well-run riding stables, horses are on preventive veterinary medicine programs. But far too many backyard horses are never seen by a veterinarian until they are sick or injured. All too often horses like Mr. Tag Along, with teeth so sharp that they can't eat properly, and so full of worms that they are skin and bones, are just considered "hard keepers." Being constantly hungry, since what nutrition they do take in is consumed by the worms, their whole personality changes. They don't want to leave the barn, where the feed is. When being ridden they keep either turning back or reaching for every blade of grass or leaf they see. Some of them never get over these habits even after they are properly cared for.

It would take a whole book just to explain what you would like to know about worms in horses, and another to tell about horses' teeth and their need for care, but for now we'll just talk about the basics.

Preventive Medicine Program

In order to be most efficient, a preventive veterinary medicine program for the backyard horse should cover three things —worming, dental care, and immunizations. It should include worming in the spring for strongyles, giving a tetanus toxoid booster along with an injection to immunize against equine encephalomylitis (sleeping sickness), and any other vaccine your veterinarian suggests for your locality and situation. In the fall the worming should include something to remove bots. In addition, immunization injection should be given for anything such as flu that your veterinarian thinks is needed, and teeth should be checked.

Worms

Horses are not born with worms, as puppies and kittens are, but start picking them up at birth either as worm eggs or larvae from recently hatched worm eggs. The eggs are passed in the feces of infected horses, including the foal's mother. Worms of the most important kind, so-called blood worms or *strongyles*, live in the intestinal tract of the horse by the million. Although tiny, not much larger than a 1- to 1½-inch strand of hair, there are so many of them that they literally suck the blood out of the horse; thus their common name "blood worm."

These worms constantly produce eggs that go out with the feces onto the ground where they are picked up by any horse that eats in that area. Back inside the horse the eggs hatch and the larvae burrow into the intestinal wall and travel through the bloodstream, doing serious damage to the blood vessels.

In young horses the larvae form of the strongyles may cause so much damage to the blood vessels that some vital arteries may become permanently damaged and then blocked, causing fatal colic or reduced ability in performance horses. After living in the blood vessels for a period of weeks, the larvae again enter the intestine, attach themselves, suck blood, and start the cycle all over by producing eggs.

Until recently, we could kill only the mature worms in the intestine with the worm remedies then available. Thus, with a horse like Mr. Tag Along, with a bloodstream full of larvae, removing the intestinal worms helped only temporarily. New worms kept coming out of his bloodstream to reinfect him. In addition, the larvae had done so much damage to his system that it literally took years to heal.

Horses are also bothered by other types of worms. *Parascarids*, the big white "spaghetti" worms, are next in importance, usually most seriously affecting foals and young horses. *Tapeworms*, said by some to be harmless, but a factor to consider in horses that are poor keepers or have chronic diarrhea, are becoming a problem in some areas. There are several other less common kinds, some of which spend their life cycles in parts of the body other than the digestive tract.

Another intestinal parasite, not actually a worm but a grub-like larvae that lives in the stomach of the horse, is the so-called *bot*. Bot flies lay eggs on horses' legs and faces. After the horse licks the eggs, the bot egg hatches in the horse's mouth or stomach, burrows in, goes through the system and, in the fall, returns to the stomach and attaches itself to the stomach wall. Although bots are terrible to see in an autopsy and were thought for years to be the cause of ruptured stomachs in horses, modern veterinary parasitologists feel they do little serious harm. Still, in the North, if your horse is wormed in late fall or early winter it is general practice to use a remedy that will kill bots. If every horse in the area is botted every fall for several years in the North, or in the South several times a year, few bot flies are seen. Just to be rid of the annoying bot flies is of great value.

Basically, if you can get rid of and keep your horses free of strongyles you can feed enough less grain to pay many times over what the worming costs. Besides, worm-free horses look better, have better hair coats, aren't apt to be itching all the time, and will live longer with far less chance of colic or other worm-related problems.

WORMING

Depend on your veterinarian to recommend a worming program best suited to your area. A fecal sample can be tested to see if horses are passing worm eggs, but it is not done routinely on all horses except to check an occasional horse that doesn't seem to be doing well, or a new horse whose worming history is unknown.

The usual practice is to worm new horses every other month for the first six months or year. In a situation where there are only a few horses and plenty of clean pasture, twice per year is often enough. Farms with large concentrations of horses often worm every four weeks.

You can buy worm remedies to use on your horse without the assistance of a veterinarian. Although most of these are good wormers, they are often wasted by being used unnecessarily or, worse yet, they aren't the correct wormer for your particular horse and situation.

Teeth

Horses have incisors top and bottom, used to bite off grass and sometimes to bite each other or unwary people. Chewing is done by a row of premolars and molars on each side, top and bottom. Most geldings and an occasional mare will have a canine tooth between the incisors and the first premolar. Even so, there is an interdental space where the bit is placed. A vestigial first premolar on the upper jaw and, in rare cases, on the lower jaw, just ahead of the first premolar that you can see (actually the second premolar), smaller than a lead pencil in diameter and sticking up only 1/4 inch or so above the gum line, is the so-called wolf tooth. These interfere with the bit on driving horses. They are usually removed at two years of age on Standardbreds, and sometimes on other breeds, particularly Thoroughbreds in training for racing. I mention these here so that you know they exist. There is much superstition surrounding wolf teeth, but it is no superstition that they occasionally cause a horse to throw its head or pull to one side to avoid the bit hitting one of these sharp little protrusions. It is a simple matter for your veterinarian to remove them with the correct tool.

Horses' teeth continue to grow through-

A HORSE'S AGE AS SHOWN BY ITS TEETH

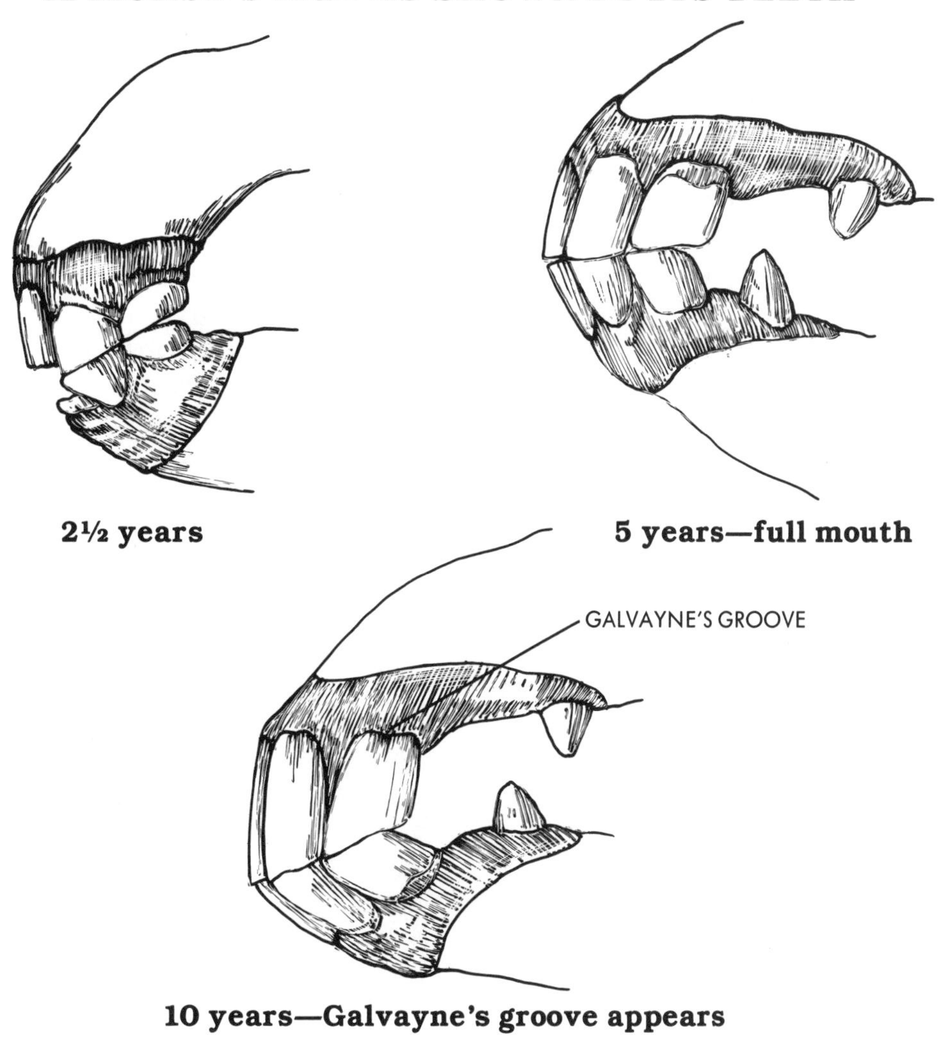

2½ years

5 years—full mouth

10 years—Galvayne's groove appears

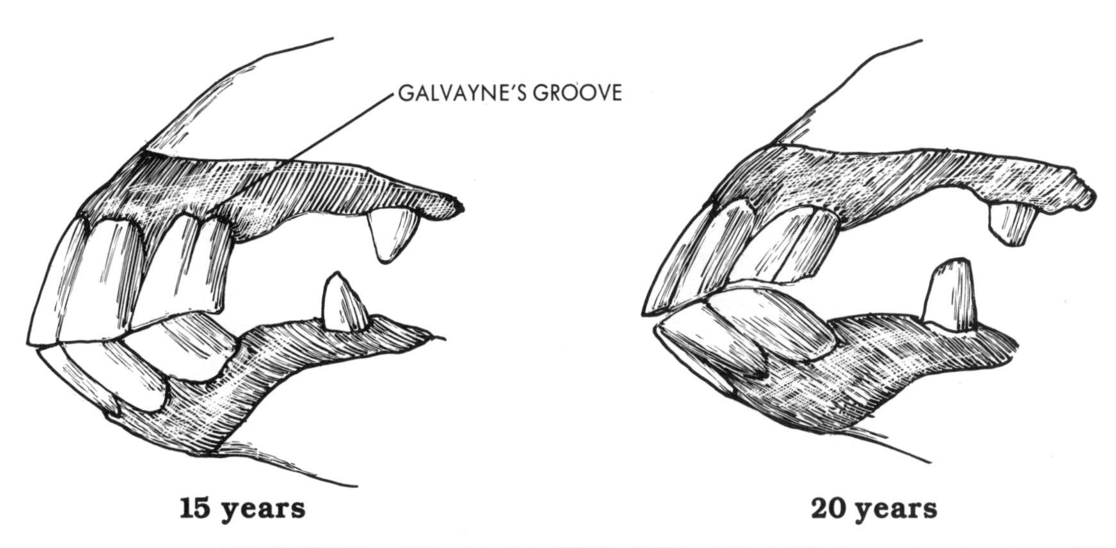

15 years

20 years

out the normal life of the horse, from eruption until the animal dies, or until the teeth wear away completely from the action of chewing. Of most concern to you as an owner of mature horses are the premolars and molars. During summer when horses are on soft green grass the molars do not wear down as fast as they grow, and they develop sharp points. These points are not on the wearing surface, but on the inner margin of the lower teeth and outer margin of the uppers. They are filed down using a long-handled instrument called a "float," which used to be rasp-like, but now is apt to be carborundum-coated. One can take off too much with a float, however, leaving a horse with teeth so smooth that grinding hay is difficult.

Many horses, from age five when there are no more baby teeth or "caps" until their mid-teens, don't need to have their teeth floated. Many owners feed a couple of hard ears of corn to each horse every day to help them keep their teeth in good shape.

Horses with malocclusions, such as parrot mouth (the top jaw longer than the lower) may need to have floating done several times a year. Other horses, during their early years, develop points on the second premolars (first large one visible) which interfere with the bit and need attention more often.

Most veterinarians in equine practice work on horses' teeth, although some farriers will float teeth and there are so-called equine dentists, many of whom are skilled professionals with years of apprenticeship. These equine dentists usually visit the larger stables and the race tracks. If you have a horse with a dental problem, and if for some reason your veterinarian doesn't work on teeth, he or she may suggest an equine dentist who is available and reliable. I have noticed quite recently, however, that veterinarians who have graduated since the mid-1980s *have* had training in equine dental work, similar to what was taught prior to the end of World War II.

A horse's age can be determined by its incisor teeth but it takes years of experience to more than just estimate, particularly after the animal has reached five years of age.

Immunizations

TETANUS

The third part of a preventive veterinary medicine program for the backyard horse is immunization against infectious disease. Most universal and most important is tetanus or lockjaw prevention as, once contracted, the disease is nearly always fatal. The organism that causes tetanus is almost everywhere, and horses are particularly susceptible to it. Under modern equine care pregnant mares are given a toxoid booster before the foal is born to give the foal immunity through the colostrum milk the first few hours of its life. By three months of age, if not sooner, the foal is given a series of tetanus toxoid injections to build up its immunity. From then on a booster is given once a year. If a horse receives a major wound, a deep wound, or a puncture wound and has not had toxoid in the past ninety days it is considered good practice to give a booster.

ENCEPHALOMYELITIS

From time to time, most of North America has had outbreaks of Eastern or Western equine encephalomyelitis (EEE and WEE), also called sleeping sickness. This nearly always fatal disease is spread to horses (and humans) by mosquitoes from birds that are carriers. It cannot be spread from horse to horse, or from horse to humans. Certain areas of North America have bird populations that are reservoirs of infection. Eastern Massachusetts,and in fact much of the East Coast, is one of these areas. Theoretically, migratory birds could carry the disease from one area to another. Ask your veterinarian if he or she recommends giving a sleeping sickness booster with the annual tetanus toxoid, and follow that recommendation. This is usually given in the spring before the mosquito season begins or, if there is an outbreak, to every horse in the area regardless of time of year.

Another form of encephalomyelitis, Venezuelan (VEE), spread from South and Central America as far north as Texas about 1970, but has since been eradicated within the United States. This, too, is spread by mosquitoes, but can also be transmitted directly from horse to man and horse to horse. Unlike EEE and WEE, however, VEE is seldom fatal in horses. Should this disease threaten in the future, horse owners and veterinarians will be warned by the United States Department of Agriculture or state health departments in time to have immunization done.

OTHER IMMUNIZATIONS

Depending on local conditions, the age of your horse, the possibility of contact with other horses, and weather conditions, your veterinarian may recommend immunization against equine influenza, rhinopneumonitis, strangles, leptospirosis, and even rabies.

Equine Influenza. Equine influenza is primarily a disease of horses less than five years of age, but can be seen at any age in those that have not been in contact with other horses during the early years of their life. It is usually a mild respiratory disease with fever, but if a horse is worked or stressed while in the early stages it can become serious, resulting in pneumonia and death. Any horse showing illness with fever, particularly when recently exposed to other horses, should be considered a possible "flu" case to be seen by a veterinarian.

Rhinopneumonitis. Rhinopneumonitis, caused by a form of flu virus, affects the upper respiratory tract. Usually mild, in certain cases it affects the nervous system, causing paralysis and death. If contracted by a pregnant mare it can cause abortion any time between five months and term. On breeding farms it affects weanling foals in a form that, because of purulent nasal discharge, is referred to as "the snots." However, it is of little concern to owners of mature non-breeding horses that are not shown or raced. If you have young horses going to shows, racetracks, or areas where horses congregate, your veterinarian may recommend vaccination against this disease.

Strangles. Strangles is a bacterial disease caused by *Strep equi*, once referred to as horse distemper or shipping fever. The initial symptom of this disease is fever such as is seen in flu, but after a few days swollen salivary glands develop under the jaw, and abscesses form. If recognized early it is sometimes treated successfully with penicillin, but there is controversy as to whether any antibiotic should be used once the abscesses form. Some strangle cases develop abscesses of the pancreas or internal lymph glands, nearly always resulting in death. Most cases, however, if nursed and not stressed, will recover with little ill effect.

What you need to remember about strangles is that, although it generally affects only young horses, a horse up to any age, if never before exposed, can contract it from diseased horses, common drinking facilities, or even from stables that had housed infected horses years earlier. Because of this, rely on your local veterinarian for advice on preventing it and always consult your veterinarian before moving horses to shows, sales, trail rides, and events where horses are brought together. To repeat, always check the temperature of sick horses, and call your veterinarian if fever above 102° is evident.

Leptospirosis. Leptospirosis is usually a mild disease in the horse, caused by bacteria. It affects most species of warm-blooded animals, including humans, with varying virulence. In the horse it is often blamed for an eye disease that comes and goes, *periodic* *opthalmia*, so-called moon blindness. Leptospirosis is an interesting disease, but usually of no concern to owners of mature horses not used for breeding and kept away from stagnant water that could be contaminated by other animals that carry the disease. If your area has a leptospirosis problem your local veterinarian is the person most apt to know about it, and will recommend immunization if it is needed.

Rabies. Horses can contract rabies if bitten by a rabid animal, most often a wild creature such as a fox, coyote, or raccoon. Vaccination is possible and is practiced in areas where the disease is rampant.

EIA

There is one disease for which there is no immunization. It seldom kills, but can cause a horse to be an invalid or, without showing any symptoms at all, to be a carrier for life. This disease, equine infectious anemia (EIA), or swamp fever, is spread by blood transfer from horse to horse. It has always been with us but only became a serious problem in the 1960s when careless use of hypodermic needles, syringes, and medication spread far more cases than did biting flies and insects. Blood-testing of all horses, quarantining or destroying reactors, and strict sanitation, including the use of disposable hypodermic syringes and needles, have nearly eliminated the disease.

Horses being raced, shown, sold, or crossing international or state lines must be blood-tested at required intervals using the Coggins test (named after Dr. Leroy Coggins, who developed it). Most states and Canada require a test at intervals of one to three years on all horses that are moved on public highways or public land. Thus, along with "teeth, worming, and shots," your veterinarian may recommend a blood sample be taken for the Coggins test.

Having personally seen the ravages of this disease in the 1960s I cannot emphasize too strongly the importance of attempting to eradicate it entirely.

You, as well as your veterinarian, should keep a written record of "shots, worming, and teeth" for your preventive veterinary medicine program to be of any value. Relying on memory is too unreliable for so important a function.

CHECKLIST FOR SHOTS, TEETH, WORMING, AND COGGINS

- **Ask your veterinarian about the appropriate worming program for your area.**
- **Have your veterinarian check your horse's teeth in the fall.**
- **Follow your veterinarian's recommendations regarding immunizations.**

CHAPTER X

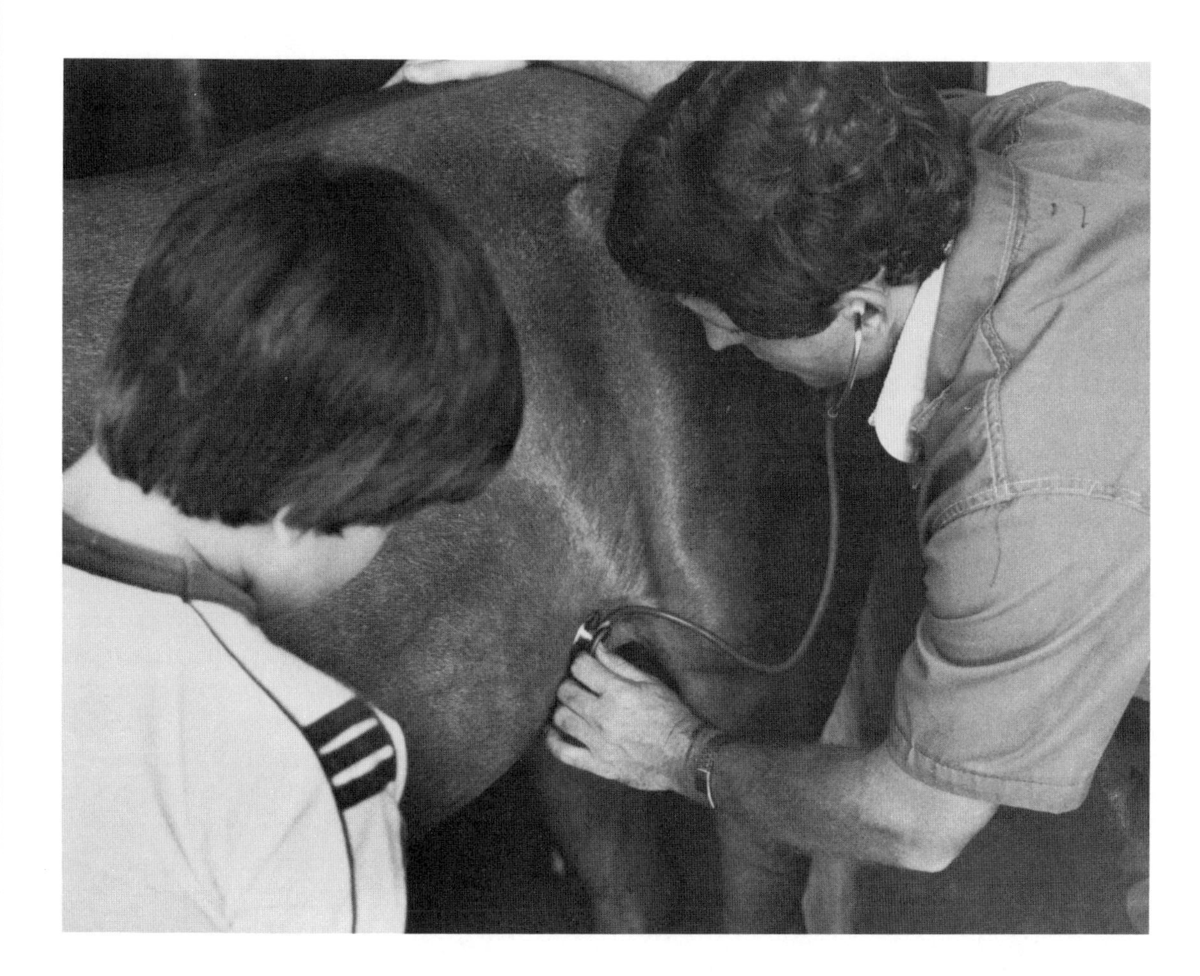

RICHARD LAKIN, MEDIA SERVICES/EQUUS MAGAZINE

COLIC, CHOKE, AND FOUNDER

*E**cho Valley Farm** was my favorite client. Two hundred well-bred Standardbreds roamed plank-fenced well-groomed pastures. The owner and his brother, fourth-generation Standardbred breeders, were working managers. One was in charge of the breeding operation and the other was in charge of the colts that were broken every fall and sent to sales or to the track every spring. The assistant managers were a young couple with life-long horse experience and college degrees in horse husbandry. The help were the pick of the horse people available in the area, taking pride in the farm and always referring to it as "the farm" or "our farm."*

During foaling and breeding season I was there seven days a week, and the rest of the year every Monday, Wednesday, and Friday, carrying out what I thought was the perfect equine herd health program. Other than foaling cases, most of which were happily completed before I arrived, there were few emergencies. Of course, there was the usual colt that needed stitches, having gone through the fence after being scared by a highway department brush chipper. But colic, the emergency that puts a cold lump in the pit of the stomach of the most experienced veterinarian, was almost unheard of among the permanent residents of the farm. When a case did occur it was in a newly arrived mare or colt with a poor worming history.

This all changed one June evening when I was called to see an acute colic. The mare had been on the farm since January and, although thin when she arrived, was now fat and sleek, already showing signs of her ninety-day pregnancy. Lady Teller had eaten her evening grain and been fine an hour before. When the night watchman arrived at the farm he found her down and rolling as though in pain. He and one of the resident farm help were trying to walk her to prevent further rolling when I arrived. They had already given her a dose of painkiller for colic.

Before I even brought my vehicle to a stop at the paddock gate I could see poor Lady was big and round with gas, wet with sweat, and attempting to lie down even with one person leading and one behind with a training whip. I didn't need thermometer and stethoscope and a peritoneal tap to know I had a serious case of tympanic colic, better known as "gas" colic. I knew I had to get a stomach tube in Lady to relieve the gas, but the usually gentle mare was so violent we were unable to do this immediately without giving more painkiller.

With this type of colic the pain is so great that the wonderful selection of sedatives, tranquilizers, and relaxants we have today is sometimes useless. Even with two good horse people to help, passing a stomach tube into this mare was like trying to stop the Dallas Cowboys with a team of three high-school boys. Suddenly the mare threw herself to the ground and went into the spasms of death.

The autopsy on Lady showed a ruptured stomach as the immediate cause of death, but what caused the sudden gas attack was unsolved. When we opened the poor mare, however, I noticed in the gas that escaped a fermented corn odor, reminding me of a mountain still I had once visited.

The next morning while I was at the farm on a routine call I noticed a second colic. This one was what we call a spasmodic colic, the form of the disease most often seen, with the pain coming from spasms of the small intestine. The painkiller calmed this mare, but with gas already forming in the stomach it was necessary to tube her and give her a gallon of mineral oil and an antiferment mixture. Again, the gas escaping from the stomach tube reminded me of my boyhood visit to the back-country still.

The next morning the best-bred mare on the farm was found dead at pasture. The night watchman had fed her at 5:00 A.M. prior to leaving at six o'clock and said all was fine. At seven o'clock when the day help arrived they found the mare dead. Appearing as though she'd lain there for forty-eight hours, she was so bloated that her upper legs pointed skyward at a 45-degree angle. However, her still-warm body showed that she had just died.

On autopsy we found the stomach ruptured, as with Lady, and again the fermented corn odor. My first question was, did you get in a new batch of grain? The answer: "Yes, but we get in a new batch every week," and "No, we haven't had them change the formula — it isn't supposed to contain more than 30 percent ground corn in the pellets, the same as we've fed for the past six years."

The grain mill was called and their answer was a surprise. They had reduced the corn and substituted barley because of the cost. I knew little about barley except that in many parts of the world where corn is not available barley is commonly fed to horses.

We put some of the pellets in a styrofoam cup, adding warm water, and put an air-tight cap on it. In fifteen minutes the cap blew off the cup. A phone call was then made to a nutritionist employed by the grain company. He claimed that barley can be substituted for corn pound for pound, but suggested we send some samples of this batch of grain to the laboratory for analysis, and not feed any more of it. He said he'd call the mill to bring in a new load before evening.

Ten minutes later he called back, asking if I had an old copy of Morrison's Feeds and Feeding. *He recalled that in his old Morrison's from college days he had read that although it is all right to feed barley to horses, it must not be ground. In order to make pelleted feed, the various ingredients are finely ground, then moistened and pressed through a machine to make pellets. He assured us that the new batch of grain would contain corn, the same as had been fed for years, and no barley.*

Why only these three animals "colicked," when it was fed to all the animals on the farm, we could only guess was that they were the best and fastest feeders, eating more than their share at a very rapid rate. Also one might guess that these three horses had something in their digestive juices, a different enzyme or even a different pH than the other horses had.

The horse is a delicate creature. No matter how much we learn and care, something new or something old we didn't learn, or Nature itself, can play tricks on us and make us realize we don't know it all.

In the last chapter we described health problems that you can prevent in your horse by routine veterinary care. Other health conditions can be prevented by good management and are often caused by an owner's lack of knowledge about the unique, delicate digestive system of the horse. We have discussed this in Chapter III, Feeding and Pasture. As a horse owner you must learn the symptoms of colic, choke, and founder, because even the most conscientious and knowledgeable horseperson can have things beyond his or her control cause digestive upsets in the horse.

Colic

Colic in the horse is actually a symptom or sign of a condition causing pain. Although usually considered a digestive upset, colic symptoms may also be caused by blood vessel blockage in the abdomen or rear legs (thrombo-embolic colic). Another cause could be mechanical blockage of the digestive tract due to *torsion* (twisting of the gut), or *intussusception* (telescoping of the gut) or, in rare cases, blockage of the urinary system.

SYMPTOMS OF COLIC

Symptoms of colic may be any or all of the following: refusing grain and hay; kicking at the belly; lying down and getting up only to go down again; attempting to roll with tack on; pawing; kicking out behind; sweating; bloating; looking at belly; squatting; squealing; straining as if trying to pass urine; banging head against the wall; biting at its side or unusual biting at humans. Rectal temperature may vary from below normal on the horse that drank too much cold water, to a slight elevation of 102° to 103°. High fever, 104° or above, and colic usually mean a fatal case.

Colic with a pulse below 45 is usually curable. Above 60 is cause for alarm, over 80 extreme alarm, and over 100 usually means a dead horse. Listening to the belly wall of a horse with colic using a stethoscope or your ear will usually reveal increased belly noise. (One should listen to many normal horses, however, before making judgment on sounds.) An absence of noise is more serious than too much. In early slow colic this generally indicates an impaction; late in a case, after a violent period, it may mean that peritonitis has already begun. The horse may even act better, but is apt to be dead in a matter of hours.

A veterinarian is always suspicious of a horse that suddenly stops "colicking," just stands there depressed, and feels cold to the hand. Coupled with a fast (80 to 120) pulse, this often indicates a ruptured stomach, which always results in a dead horse.

There are three basic types of colic:

1. *Spasmodic*, caused by violent contractions of the gut (usually small intestine). This form is usually acute. It may stop as quickly as it starts, or it may be intermittent.

2. *Tympanic*, or gas colic, caused by excess gas accumulation in the stomach or other portions of the digestive tract. This form, too, is usually acute and violent. It often leads to fatal torsion or may result in rupture of the stomach.

3. *Impaction*, caused by blockage of the gut, usually the large intestine, by ingesta. This form is usually a mild "slow" colic, but seldom gets better spontaneously without vigorous treatment.

Any colic can suddenly become violent and fatal if a horse rolls while "colicking," allowing portions of the digestive tract to twist and, in doing so, cord off blood vessels. It is possible for a horse to be eating and healthy, but kick out a hind leg at another horse and from the exertion cause a section of large intestine or cecum to twist. There are also occasional cases in the horse of intussusception, or telescoped gut, causing violent and usually always fatal colic. These twists and intussusception can only be cured by immediate diagnosis and surgery before rupture.

Peritonitis is acute bacterial inflammation of the lining of the abdominal cavity. It can be caused by the rupture of any part of the digestive tract. A horse will die of peritonitis within hours, so the success rate of colic surgery is low, even when diagnosed early and operated on immediately.

The greatest fear any veterinarian has is that a mild slow colic will suddenly turn into an acute, violent, fatal colic because the horse rolls on the ground, causing its intestines to twist.

Influence of Parasites. Veterinarians used to see many cases of thrombo-embolic colic caused by blockage of blood vessels damaged by strongyle larvae. Many cases of simple spasmodic colic result from worm larva lesions blocking the small blood vessels that supply the intestine. It has been written, and from experience I agree, that at one time most of the colics we saw were parasite-larvae-induced. With better worming practices this is no longer the case.

"Kidney Colic". One will often hear of a fourth type of colic referred to as kidney colic. Although caused by problems with the urinary system and seen in cattle, sheep, and dogs, it is rare in horses. Horses with severe colic will often appear to be trying to urinate, giving rise to the term kidney colic, in the belief that the urinary tract is the cause. Young veterinarians were often advised by their more experienced colleagues that it was easier to agree with the owner that the horse had kidney colic, administer a diuretic so the horse would pass urine, and treat the horse for whatever really ailed it, than to argue with an owner old enough to be your grandfather. Further, if the horse got better everyone was happy, and if it didn't, "that young vet tried, but even though he got him passing water he was too far gone."

TREATMENT

You should call for professional help at the first indication of colic. Despite all we have to work with today, some horses with colic will die, some even from self-inflicted injury. Don't be surprised if your veterinarian asks you to stay with a colicky horse until recovery is indicated. Nursing care, preventing the horse from rolling, and protecting it from injuring itself are vital to recovery.

Often symptoms of a mild case of colic can be alleviated with a liquid medication such as Maalox left by your veterinarian for this purpose.

Choke

A horse that suddenly stops eating after a few mouthfuls, or is found straining and squealing, head and neck extended, with food particles coming from its nostrils, is suffering from *dry choke*. Choke is when a wad of food materials blocks the esophagus. This is often confused with colic. Should one make this mistake and attempt to give liquid medicine with a dose syringe, some of the medication could end up in the lungs because the horse is unable to swallow Always keep choke in mind before you decide to use liquid by mouth to treat colic.

CAUSES

The causes of choke are many. Horses that are dehydrated from racing or work, and are fed fine hay, such as third-cutting alfalfa, beet pulp, or finely chopped hay or dry grain, prior to being allowed a little water, are good prospects for choke.

TREATMENT

Many cases of choke will correct themselves if left quietly alone. Others will require all the patience and skill your veterinarian can muster. You should call your veterinarian at the first indication of choke, but despite the disease's frightening symptoms patience is the main thing required.

Founder

After-effects of colic caused by overeating, or overeating grain without colic symptoms, may cause *founder*, or *acute laminitis*. Laminitis is inflammation of the inner lining of the wall of the hoof, causing it to become engorged with blood, literally separating the wall from the inner tissues of the foot. It is caused by any toxic agent a horse may eat — particularly the toxic products of fermenting foodstuffs — or by too much grain, particularly grain meant for chickens, hogs, and cattle. Retained placenta, high fever, and/or digestive upsets, if not properly cared for, or even if properly cared for, sometimes result in founder. Many cases of Potomac Fever, a recently recognized disease that causes fever and acute diarrhea, end up as acute cases of founder.

Horses of draft type and the smaller heavy pony breeds are especially susceptible to founder, to the point of coming down with it when pastured on lush green grass. And certain light horses, such as heavy round Quarter horses and Morgans, seem more susceptible to it than are fine-boned Thoroughbreds and Standardbreds.

Sick horses should have their feet checked several times a day for any sign of heat, usually the first symptom of founder. Once the condition sets in the horse will appear frozen to the ground. A horse carries 60 percent of its weight on the front feet and they are usually the first to founder. The typical foundered horse carries his rear legs up under him and appears to be sore behind. The fact is that he is attempting to get the weight off the front feet.

TREATMENT

First aid for founder is standing the horse in cold water, an old but still effective remedy. On even the slightest suspicion of founder, call your veterinarian and follow directions. Many treatments can be successful if the condition is recognized early. However, founder resulting from Potomac Fever or from retained placenta (holding the afterbirth) is usually irreversible.

Ponies and horses with ski-shaped feet are suffering from chronic founder (chronic laminitis). They can be treated and brought back to some use by skilled farriers, but often the treatment seems worse than the disease, causing such damage that the horse must be euthanized. Most important, be aware of the danger of the acute form and do everything you can to prevent it.

COOLING OUT

One of the major causes of founder or acute laminitis is neglect in the cooling-out process. Although this is mentioned on page 44 it might well be described in more detail. When returning from a ride or drive during which your horse has been sweating, you should walk the horse on a lead until active sweating is stopped. If the weather is cold, or there is a cool breeze, a blanket or light sheet should be used during the cooling-out process.

During the cooling offer the horse water, but allow him only five to eight swallows every ten minutes. When he is cool enough that he won't take any more he may be allowed free access to water. A variation of cooling is to wash the horse down with lukewarm water, dry him with a sweat scraper, rub him with towels, and put a blanket (or "cooler") on him. He is then walked or tied in his stall or cross-ties. Regardless of which system or combination is used, *do not feed grain to a hot horse.* Coarse hay may be fed any time, but fine hay can cause a hot dry horse to choke.

Once a horse feels cool and dry to the touch of your hand under the cooler, remove it and either put a light sheet on the horse, or if he is used to no cover leave him uncovered in his stall or paddock. If during the cooling process he would not drink water be sure he does not gorge on it when he returns to his stall.

CHAPTER XI

STEVENS

TRANSPORTING YOUR HORSE

__Treasure__ was well named. A dark bay Thoroughbred mare, with a racing background, excellent disposition, clean legs, and the will and heart to jump, she certainly was a treasure. Contrary to popular belief, the Thoroughbred racehorse is one of the easiest breeds to work with. They are trained by professionals, from being haltered at birth to being ridden at eighteen months, and are accustomed to the noise and commotion of the racetrack at two years. Typical of this class of horses, Treasure would walk up the ramp of a horse van or into a horse trailer with the same obedience and willingness with which she walked into her stall.

Treasure's owner was a young woman with an excellent 4-H and Pony Club background, having learned to ride at an early age under professional instruction. She bought Treasure with the idea of using her for more advanced showing and jumping, and secondarily as an investment as a brood mare. Treasure and Jody, her owner, were a winning combination at show after show. Jody's family had their own horse trailer, using it to transport Treasure and their other horses to shows. Before loading, Jody would always wrap Treasure's legs from just below the knee or hock to below the coronary band. This was done to prevent injury to the most delicate and most vital part of the horse's anatomy, the lower leg.

For two years Treasure was transported to and from shows without incident. Then one day when she was unloaded at the show Jody noticed that hair was rubbed off the point of her hock on one side. The mare was sweating, seemed nervous, and did not do well at the show. Upon returning home, both hocks showed sign of injury. Checking the trailer door, Jody saw signs that Treasure had been kicking. A week later her hocks were bandaged prior to leaving for a show, and on arrival it was found that she had kicked so hard that the bandages came off. This time she was reluctant to load for the trip home, not unusual for some horses, but for Treasure a first, and for Jody an embarrassment.

Because of her scarred-up hocks, Treasure was not shown for a few weeks. Then, hoping she would not act up again, Jody loaded her for an important nearby show. Quite by coincidence, one of the local 4-H leaders followed the trailer hauling Treasure. Arriving at the show, she reported that each time the trailer brake light came on she could see Treasure kicking. For the return trip home Jody disconnected the plug to the electric brake and trailer lights. The same woman followed and observed this time that although Treasure had again been reluctant to load, she rode all the way home with no visible signs of kicking.

A complete rewiring of the electric system of the trailer, replacing old wires having worn spots and cracked insulators, re-attaching grounds and replacing sockets, solved the problem. Apparently Treasure had received a shock similar to that of an electric stock prod from the metal sides of the trailer each time the electric brakes were applied.

Red was similar to Treasure in that she, too, was purchased by an experienced young horsewoman to be trained to be a show jumper. She was different, however, in that she was purchased as an unbroken two-year-old. Although supposed to be of Appaloosa ancestry, she was a solid red chestnut. Red's owner was experienced and talented in training beyond her years, having had qualified professional instruction in riding and experience with a professional trainer.

Red was first saddled at three years of age and took to training so well that at four she was jumping anything she was aimed at with grace and beauty. Loading her on a trailer, however, was a different story. Each time Red was to be shown her owner didn't know whether the mare was going to go into the trailer or not. Sometimes she would walk on with no problem; other times it took all the patience one could muster, and finally several people would get behind her and literally lift her on.

One day Red pulled back after being loaded, her leather halter started to break and a sharp rivet cut into the facial vein. Blood poured out and it took stitches to stop the bleeding, but not before the wooden floor of the trailer got so slippery that both Red and the other horse in the trailer became terrified as they lost their footing. After that Red never loaded easily — until Clarence Martin took over.

Clarence was a legend. He had shod horses from boyhood, was a farrier with the

U.S. Expeditionary Forces in France in World War I, and trained numerous farriers as apprentices. But most important, he was an expert on whip training, a method he had learned from German P.O.W.s while overseas. Whip training does not involve whipping a horse, as one might think, but teaching the horse by using the whip almost as a wand.

Story after story is told of Clarence appearing at a horse show and taking over a horse that others had tried for hours to load. When left alone with the horse he would have it on the van or trailer in minutes. He was consulted on the problem of loading Red. With his usual quiet manner and love of teaching humans as well as horses, he took over. The last time I ever saw Red load, like Treasure she went on the trailer with as much willingness as if she were walking into her own stall.

Assuming that your horse arrived at your place aboard some sort of motor transport, you know that it can be loaded. Of course, you might be lucky enough to be near trails, or perhaps you have enough land of your own so that you never have to transport your horse again. Even so, sooner or later you may wish to participate in a show, trail ride, or other event. Being able to load your horse onto a trailer or van with no more fuss than if you were leading it into a stall adds a great deal to its value. On the other hand, if you have acquired a horse that absolutely refuses to load or requires five people to literally carry it onto the transport vehicle, the pleasure you receive from the show, trail ride, or other gathering is eliminated before it starts.

If you plan to transport your horse, try a dry run first. Have the transport vehicle you plan to use brought to your place a few days before the actual event. Load the horse, drive a short distance, return, and unload. Some horses load easily on vans and won't go into trailers, or vice versa. Don't ever try to load a horse onto a horse trailer that is not securely hitched to the pulling vehicle. The trailer will tip back as the horse puts its weight in the rear, frightening the horse so much that it will never load easily again.

Loading

If you find that your horse will not load, inquire around until you find a trainer who is experienced in breaking horses to load. You will get advice on loading from everyone you talk to — most of it sound for someone else's horse, but for some reason it won't work on yours. Paying a trainer who knows the business of breaking a horse to load is as important as paying someone to break your horse or to teach you to ride.

PREPARING HORSE AND TRAILER

Before loading a horse on any vehicle for transport, all four legs should be wrapped from below the knee or hock to below the coronary band. Practice this wrapping until it can be done easily with bandages staying in place. Wraps and pads for the purpose are available with the wraps secured by Velcro. Most professionals use safety pins and/or tape to reinforce the Velcro holders. (See

Chapter VIII for some information on the proper wraps.)

Don't use your best leather halter on your horse while transporting. Even under the best conditions a horse may fly back and, if tied, ruin the halter. Nylon halters won't break, but can be cut loose in an emergency. Never use a rope halter on a horse being transported. It can get too tight if wet and do serious damage to the horse. When horses are tied in a van for hours or days on a long trip, padding halters with flannel wraps or deer hide will prevent wearing hair off over the nose and under the jaw.

Some people will advise you to use tranquilizers on horses prior to loading, but in most cases they are not needed, and in competitive meets they are illegal. Some high-quality but coarse hay in a hay bag is all the tranquilizer most horses need while being transported. An exception might be a high-strung young horse being transported for the first time. Unless you have had problems with a horse and can't find the reason, don't use tranquilizers.

One of the most common reasons for horses being upset on trailers is stray electricity from lights or electric brakes. Wiring in trailers is easily damaged and often neglected. Once frightened by electricity a horse never forgets it, so stay away from dilapidated, poorly maintained equipment when renting or buying transportation for your horse. If you own your own trailer, make sure you check the wiring before every trip.

For years it was recommended that horses being transported long distances, more than eight hours on the vehicle, be "oiled," that is, given a gallon of mineral oil by stomach tube just before loading. This is done to prevent the danger of impaction due to dehydration while being transported. A better method is to give a bran mash in the feeding prior to loading and, of course, to provide plenty of water at every rest stop. The first feeding after unloading should also be a bran mash (see Chapter III).

Buying a Trailer

If you are going to do a considerable amount of showing, trail riding, or transporting your horse for any reason, consider owning your trailer. Second-hand trailers are often available at very reasonable prices. On any second-hand trailer, or one you own for more than a year or two, check the wiring and floor boards and, if they are deteriorating, replace them rather than repairing them. Hitches and safety chains should be checked regularly by someone who is familiar with horse trailers. Don't forget to check the tires before every trip. Remember, you are handling a load almost as heavy as your vehicle, and safety to you, your horse, and other motorists cannot be compromised.

Beware of low trailers meant for ponies, small horses, or cattle. Any horse, even of small size, needs room to raise its head without touching the roof. Also beware of home-built or -rebuilt bargain trailers. Proper non-slip rubber or fiber mats should cover the floor. A little bedding will encourage a shy horse to walk onto the trailer, but can cause slipping. When available, peanut hulls make a good non-slip bedding for trailers, and the horse can't reach down to eat them.

There is a great personal satisfaction in being able to load your own horse on your own trailer and take it wherever you want to go, increasing your pleasure in owning a horse.

The vehicle you use to pull the trailer should be of sufficient size, power, and weight to handle the weight of the trailer and two horses, not only for uphill, but for control going downhill as well. Trailers themselves have brakes controlled from the tow vehicles. If these should fail, the tow vehicle must be heavy enough to stop both. Special "trailer packages" include heavy-duty springs, shocks, and hitches and are available not only for new vehicles but may also be installed on good used vehicles.

A four-wheel-drive vehicle is a great advantage in moving a trailer in and out of wet fields used as horse show parking areas. Since many autos today are too small and light to safely handle a horse trailer you might find you have to go to a pickup truck for a source of power.

Practice driving with the empty trailer before you load horses in it to get used to starting, turning, and stopping. Learn to anticipate stops, slowing down gradually, using trailer brakes before you apply vehicle brakes. When turning, particularly at 90 degrees, always remember you need more room to make your swing. Practice backing up in an open field or a parking lot with no vehicles around until you can do so without having to think about which way to turn the steering wheel.

CHAPTER XII

INTERNATIONAL ARABIAN HORSE ASSOCIATION

OWNING A HORSE WITHOUT A STABLE

__Meg and Don Ferris__ lived a life that would tire most people just thinking about it. An executive with a publishing company, Meg traveled coast to coast to sales meetings and conferences with the same regularity as most of us go into town to shop. Don, an executive with an electronics firm, traveled worldwide just as frequently.

The Ferrises had a home in the country where they spent weekends caring for their lawn and working a large garden. They enjoyed their lifestyle, but felt that two things were lacking. One was a sound investment policy. Second, since they were away from home so much, they could not own any pets. During her girlhood Meg had a horse and wished she could have one now.

One Friday afternoon Don's plane had to land at an airport 300 miles from home due to weather conditions. Rather than wait for his home airport to reopen and book another flight, he decided to rent a car and drive home. As is often the case in such a situation, not enough cars were available, so he shared a car with two men going to the same upstate area where his home was located. During the long drive Don found out that his traveling companions were a horse trainer and a horse owner on their way back from a Standardbred yearling sale.

Don learned from his two new friends that most racehorse owners never keep their horses at home, but instead at training stables and/or horse farms that specialize in handling "turnouts" or breeding. He also learned from the owner that although only a small percentage of owners make a profit on their horses, people who select good trainers and buy good stock find horses as good an investment as playing the stock market. The owner compared the good trainer to a good stockbroker, and noted that if you can afford to lose a bit of money you stand a chance to make some, too, not only in race winnings, but by selling a winning horse, breeding a mare and selling her offspring, or taking advantage of certain tax breaks that many states allow horse owners. He also mentioned that even if he lost some money, the pleasure he and his wife derived from owning a racehorse was more than he had ever imagined.

"On weekends and days off I go to the stable and jog my horses for their daily workouts," the owner said. "It is more relaxing than going skiing and I don't have to spend money for lift tickets."

Don asked the two men how the financial arrangements were worked out between the trainer and owner. They explained that it varied from the trainer working on a salary for the owner, as might be the case if an owner had his own farm, training track, and several horses, to the trainer owning a share in the horse and sharing expenses as well as earnings and eventual sale income. The more usual arrangement, they said, was somewhere in between with the owner owning the horse outright along with harness and equipment, and paying the trainer on a per-day basis for training and costs, according to a contract.

They told Don that Thoroughbreds (known as runners or flat racers) work on somewhat the same basis. With Standardbreds, however, people who enjoy taking part in training their own horse, even in a small way, can participate, because most people can learn to jog a horse. Some even go to the track and help "paddock" (harness and prepare for racing) their own horse on race days or evenings. A Thoroughbred racehorse owner has little opportunity to participate except perhaps to be present at pre-race saddling, and to accept the trophy if he or she has a winner.

The trainer added that he knew other trainers who worked with riding horses, hunter jumpers, and "fine driving horses," all owned by people who didn't have their own stable.

When Don became interested and asked more questions about becoming a horse owner, the trainer told him that, although he had all the "colts" he could handle that fall and winter, he knew of a reputable trainer within a few miles of Don and Meg's country home who had a couple of empty stalls and would be a good person for an inexperienced owner to engage.

As they parted company at the end of

the trip the owner said, "A draft horse owner once gave me the best reason for investing in horses. You can't go down to the bank and ask to look at your money, but you can go to the stable and look at your investment, pet its nose and feed it a carrot, or drive it in the Fourth of July parade."

When Don reached home and told Meg of his experience and what he had learned, she, too, became interested. The following day they went to talk to Paul Masters, a local businessman who owned a few Standardbreds he trained himself. Masters explained that he had gotten into Standardbreds years ago, and kept them on his dairy farm. He had sold the dairy herd when it became difficult to make a profit, but kept increasing the number of horses since they were a great source of relaxation for him. One of his former dairy farm employees took care of the horses, but he himself went to the barn every morning at daylight and hitched and jogged horses until time for breakfast. Then he put on a business suit and went to his office. During the spring, summer and fall he raced one of his horses at a track eighty miles away. He trailered the horse to the track, paddocked and warmed the horse, but hired a professional driver for the actual race. During the county fair racing season he entered horses at several fairs and drove them himself.

When Don asked him about the trainer he had heard about, Mr. Masters had high praise for him, but he warned Don that unless he was prepared to lose $10,000 or more the first year he shouldn't consider a racehorse investment.

Masters told the Ferrises that local Standardbred owners and trainers belonged to a driving club which met monthly. He invited them to attend the next meeting if they wanted to meet and talk with more Standardbred people. He explained that, as they had already learned, more racehorse owners use professional trainers than train the animals themselves, even if they have facilities at home to care for horses.

The next weekend the Ferrises went to the driving club meeting as Paul Masters' guests and met not only Bill Clark, the trainer they had been advised to see, but an assortment of people who, although they were from varied backgrounds, from bankers to day laborers, had one thing in common — they owned, drove, or trained Standardbred horses. Meg, who had been exposed to Standardbred people as a girl, had to explain to Don what they were talking about. She told Don that she knew that each type of horse enthusiast had his own lingo.

Don overheard a man say, "He winned in three after the other horse spit the bit." Meg explained that if the man had been talking to someone not familiar with harness racing he would have said, "My horse won the race at two minutes and three seconds after the lead horse quit and dropped back quickly."

Bill Clark, the trainer they had wanted to meet, invited them to come to his stable Sunday morning to talk. When they arrived they saw horses being jogged by a variety of drivers, from a twelve-year-old girl to a man who appeared to be over eighty. The trainer explained that owners liked to jog their own horses and, since he could count on them coming on weekends to do so, he could give his regular help the days off and still keep the horses on a seven-day week.

The trainer asked Don and Meg if they had any experience with horses. Don said no, but Meg said as a young girl she rode a horse and drove a pony. Clark asked Meg if she'd like to try taking a horse around the half-mile training track a few times. Although she was a bit apprehensive at first, the steady beat of the trotter's hooves, his response to her hand on the lines, and the feeling that she was in control, yet part of the horse, gave her a feeling she had not enjoyed since she rode her horse in childhood.

The following weekend, after consulting with their accountant and their attorney, the Ferrises asked Clark to find them a yearling. Two weeks later he phoned to tell them that he had found a suitable "colt." When Don said, "But we wanted a filly," Bill laughed and said, "We call the yearlings we are breaking 'colts' regardless of sex. After the first of January when they reach their official birthdays they are known as two-year-old colts or fillies." He added, "I found this one at a local farm. She's well bred by a top young sire and I raced her dam as a two-year-old until we were offered a fabulous price for her." Almost as an afterthought he said, "If you like her and think she is worth $9,000 you can make arrangements to buy her and I'll have an owner/trainer contract ready for you to sign."

The Ferrises went to look at the filly, a dark bay with an already long but shiny winter coat and a typical large Standardbred head. Comparing her to Thoroughbreds of the same age that they had seen at a nearby farm, she was small and almost homely. But recalling a painting of the great Standardbred Dan Patch, which hung in a local inn, Don realized she was typical of her breed. When Bill Clark's teenage daughter brought her out and put her in the cross-ties for Don and Meg to have a better look they realized that her gentle eyes and calm, deliberate movement gave the filly an inner beauty they had not seen in other horses. Her name, Summer Breeze, seemed to fit her well. They stood there looking at her, imagining what it would be like to stand by her head as she was blanketed as the winner of a big race.

Although it was almost "love at first sight" for the Ferrises, and they could well afford the asking price of $9,000, Meg suggested that they have their lawyer check the proposed trainer/owner contract and have the filly checked by a veterinarian other than the one who did work for both the seller and the trainer. This was when I first met the Ferrises.

A pre-purchase exam on a yearling prospective racehorse can vary all the way from a casual five-minute exam to a battery of tests, from radiology to blood chemistry. However, the history of the yearling is most important. I trusted this particular trainer from past experience. He would be apt to note conformation faults that would cause poor racing performance or breakdown in the future. Still I went over her from nose to tail, using sense of touch and sight.

All signs, from heart and respiratory sounds to reflexes seemed normal. My fingers could detect no old injuries on her legs or hooves. Hoof development and general conformation appeared normal. Tooth development and bite were normal, although there was indication that wolf teeth would need to be removed by spring. Her eyes appeared normal to ophthalmoscope examination and there was no indication of skin disease. An examination of her genital tract by rectal palpation showed her to be apparently developing normally.

Since a horse can't open its mouth wide enough and say "ah" to let the examiner see the pharynx and larynx, examination is done with a fiberoptic scope passed through the nose. This filly showed a few spots of redness in the larynx from a bout of flu the previous spring, but within the normal range. Horses do not have tonsils as most other animals do, but instead have lymphoid tissue similar to tonsils in small patches in the larynx. In too many cases an attack of respiratory disease in a young horse will leave these patches greatly enlarged (lymphoid hyperplasie) so that they interfere with breathing.

Radiographs (X-rays) are usually taken to detect weakness or lesions of the navicular area of expensive horses destined for hunter-jumper competition, but are seldom done on young Standardbred racehorses unless other visual and pres-

sure testing exams indicate a possible problem. In yearling racehorses, however, many veterinarians recommend that radiographs be taken of the radius where it attaches to the knee. In horsemen's terms this is "X-raying the knees." What we are actually doing is using a view of the growth plate area (ephyseal joint) to determine maturity and hardening of the bone. If taken in November or December, and again in February and April, one can tell if a two-year-old is mature enough to go on training and racing. These were done on the Ferris filly.

Most important in the pre-purchase exam of a yearling racehorse is the history of the animal. In this case a life history of immunization, worming, and health accompanied the filly from the breeding farm. She had papers showing a negative Coggins test for equine infectious anemia taken a few weeks before. Still I suggested a repeat of this in thirty days and booster injections for rhinopneumonitis.

When I gave the Ferrises my report I emphasized that this exam was done to the best of my ability and told only the condition of the filly that day. A satisfactory pre-purchase exam is no guarantee that a horse won't break down two weeks later or, for that matter, drop dead of a heart attack an hour later.

A day later the Ferrises bought Summer Breeze and joined the thousands of people who own horses but have neither facilities nor time to care for them.

Most other chapters in this book have started off with a story about mistakes a horse owner has made and ended with an explanation of how to avoid them. Being business-oriented, considering their horse primarily as an investment, and realizing that they knew little about what they were getting into, the Ferrises sought the advice of professionals and experts. They were warned and understood from the beginning that only a small percentage of racehorse owners make money.

Had this chapter continued in story form it would have told how Summer Breeze raced for three years, earning her purchase price, training fees, and veterinary and farrier expenses, with a great deal to spare. The Ferrises refused several quite substantial offers for her, deciding that they would retire her as a five-year-old and breed her. The economics of such a decision are controversial, but with the mare's race record and blood lines a foal from Summer Breeze by the right sire would be a valuable asset to hold, train and race, or sell.

In September of their mare's fourth year of racing she won her last race and was retired. She was moved to the Paul Masters farm where she could be "let down," turned to pasture with other mares, and prepared for breeding the following spring.

On the advice of their attorney and accountant, the Ferrises kept a life insurance policy on Summer Breeze while she was being raced. Since this was quite expensive to carry and her actual value became less as an unproven brood mare, the policy was dropped the day after her last race. This proved to be a mistake two months later when the Masters barn burned, taking the lives of five horses, including Summer Breeze. To Don and Meg, however, the financial loss was minor compared to the heartbreak of the abrupt end of their dream.

There is an awful lot of luck, good and bad, involved in the horse business as well as other businesses dealing with livestock. No matter how well we plan and manage, unforeseen things happen. Still, owning a horse with no facilities or time to care for it can be rewarding.

This type of ownership is not restricted to racehorses, and if you own a racehorse you don't have to race it. There are rewards both in pleasure and profit in ownership of a brood mare of Thoroughbred, Standardbred, and racing Quarterhorse breeds. There are opportunities to buy good brood mares of the racing breeds, and of show stock in Arabian, Saddlebred, Morgan, Hackney, and several other breeds that are considered good investment. Many large breeding farms and owners of small horse farms own only a few of their brood mares, but rely on year-round boarders to pay the bills. The profit on such ownership, if there is any, comes from the sale of offspring of these mares.

In addition to brood mares, one may buy a hunter-jumper or other show horse simply as an investment and hire others to train and show it, just as one would with a racehorse. To make any financial gain on one of these investments one must be willing to spend enough to increase the horse's value by showing and winning and then, when the right opportunity to sell comes along, to let the animal go. To be practical, one must realize that the greatest reward from owning such a horse is to see it win, and if a financial gain is possible so much the better.

In the past, owners of valuable show horses donated them to school stables, taking a tax write-off as one would for any donation. Under new tax regulations this is apparently more difficult to do.

Another type of horse ownership is in stallion shares, usually restricted to the racehorse breeds. Briefly, this involves buying one or more shares in a syndicate that buys a valuable racehorse on its retirement from racing with the hope that his breeding fees will pay a profit. Even before the recent tax changes this was a risky business, and even a person knowledgeable in the racehorse and investment business stood a greater chance of losing his or her investment and a lot more besides. Still, if a stallion's offspring prove to be winners and the stud fee goes up instead of down, those who do make a profit make a good one.

Before going into an investment in horses one must be aware of three things. First, if you can't afford to lose your initial investment, plus a lot more, don't even consider it. Second, remember that it takes more than hay, grain, and shelter to keep a horse. The farrier, the veterinarian, the insurance agent, the tack salesman, and the repairman all must be paid for never-ending services. The sad truth is that a winning horse has very small veterinary bills, but a losing horse with no income usually has high veterinary bills. Third, if you are into horses strictly as an investment, don't fall in love with any of them.

My father once said, "If someone offers you a good price for a horse or a cow, take them up on it before something happens to the animal."

Years later, when asked why he never followed his own advice on this matter, he said, "No one ever offered to buy the ones I didn't like. Trouble was, I never owned one I didn't like."

PART TWO TITLE PAGE PHOTOGRAPH BY CONTEMPORARY IMAGES, COURTESY OF THE APPALOOSA HORSE CLUB.

PART TWO

GETTING MORE ADVANCED

CHAPTER XIII

APPALOOSA HORSE CLUB

ADDING ANOTHER HORSE

*The **Howards** bought their first horse when they moved to the country and opened a glass installation business. Although they were inexperienced, unlike the Deanes they learned all they could before bringing the first horse home.*

Their first horse was a young Standardbred race filly named Second Story, stabled and trained by a local trainer. Both Sally and Steve Howard learned from the trainer how to drive the filly at her daily jogging, and found these early morning sessions great relaxation from the tensions of their fast-growing business.

After the first summer of training and racing, Story was brought home to the Howards' place where she had a well-built, warm, but adequately ventilated stable with a large box stall, and two acres of pasture-paddock where she exercised.

The Howards had a twelve-year-old son, Joe, who enjoyed caring for Story but wanted a horse he could ride. He joined the local Pony Club and learned to care for horses while he took lessons on the horses at a nearby riding school. The next step, of course, was for Joe to acquire his own horse. Being a tall boy for his age, he needed a horse, not a pony. Joe had been riding a particular Thoroughbred at the riding school and became attached to it. It seemed the boy and Ebony, the six-year-old gelding, were meant for each other.

The second summer the Howards owned Story she was with the trainer again, in the money and paying her way, even winning an occasional race. Ebony was brought home from the riding stable and a second box stall was added in the Howards' barn so there would be room for Story when she returned home in the fall. Steve helped Joe build jumps in a nearby field where the boy and Ebony worked out each day.

On a Saturday evening in early November I got an excited call from Sally Howard. "Ebony has a cut on his gaskin and we can't stop the bleeding. Please come."

When I arrived I found that Joe had stopped the bleeding by holding a gauze pad over what was actually a deep puncture wound made by an extension on Story's right hind shoe called the trailer. I cleaned and sutured the wound and gave Ebony a tetanus booster. A few other minor abrasions and wounds on poor Ebony had already been adequately sprayed with furazone powder by Joe.

What had happened was all too evident. Story, home from the races, had been turned into the paddock with Ebony. Within seconds the gelding was trying to make friends, and Story, half his size, had whirled, kicking with both hind feet, and cornered Ebony. Up until now the Howards had done everything right, but turning two strange horses together, particularly one with new shoes on its rear feet, was a serious mistake. (After proper acclimating, Story and Ebony became good friends, which created a new problem. It was difficult to remove one without the other becoming upset.)

Adding a new horse is usually a painless procedure, but it can be a disaster. Never turn two strange horses together into a paddock until they have had a few days to get to know each other over a stall wall. It may be necessary to lead them out one at a time past the other's stall. For example, you might let one use the paddock and the other stay in, and then switch. If you have adjoining board-fence paddocks turn both out so they can see and smell each other over the fence.

Most important, when turning two strange horses together for the first time be sure both are barefoot behind. Particularly dangerous are shoes with a rear projection, or trailer, often used on Standardbreds, or those with caulks, pointed projections on the bottom. Have your farrier show you how to remove a shoe by rasping off the clinches and lifting the shoe with nippers, so that you can do this when necessary.

The more space available, the safer it is to put strange horses together. Never leave machinery, brush piles, or other hazards in a paddock or pasture.

Mares aren't usually as aggressive as geldings, but there are no rules one can go by, such as that mares always get along with each other or geldings hate each other, or whatever. Even mares and geldings that are used to each other sometimes get into kicking matches when the mare is coming into or going out of heat, or, apparently, when one or the other just happens to have a "bad day." Some horses never get along, and others are totally unconcerned when new ones are added. Usually there is a period of posturing and testing. Soon one or the other will become the dominant one, most likely the mare, and peace will reign.

It should go without saying that two stallions must never be turned out together. A stallion should never be paddocked near other horses unless you have exceptionally good, thick, tight plank fencing. Although most male racehorses are stallions, and handle as well as geldings, stallion care and management is a field of horse husbandry all its own. The amateur owner should never undertake the care of a stallion without professional advice.

Adding a third horse can sometimes create more problems than putting two together. For some reason it seems that two will always gang up on the third one. This may only be a problem at feeding time, or at the watering tub. Each individual circumstance must be dealt with as it occurs. For example, if it is a problem at feeding, you may need only to tie the dominant horse.

When larger groups of horses are pastured together all should be barefoot if at all possible until all signs of kicking each other are gone. Sometimes one particular horse will be of a disposition that makes it impossible ever to let it be with others. Other times two particular horses continue to fight, but if split up and put with others they get along fine. Learning all this takes time and patience.

In any group of horses certain ones will pair off and become buddies. If you remove any two individuals from a group and place them with another large group these two will hang together. Again, if you remove one of them and another member of the second group, and put these two with a third group, they will stay together.

At breeding farms, where numbers of mares are brought together once a year for breeding, certain individuals will pair off and become inseparable. Separate them after breeding and a year or two later put them back in a group of horses and within minutes they will find each other. Horses do remember each other.

CHAPTER XIV

RICHARD LAKIN, MEDIA SERVICES/EQUUS MAGAZINE

An Experienced Horseman Buys a Horse

Joe Howard, *son of Sally and Steve Howard, turned out to be an unusual boy. There was nothing unusual about him at age twelve when he began to ride, nor at thirteen when his parents bought him his own horse. What was unusual about Joe was that when he reached sixteen he was still riding and showing. Most boys his age, and too many girls, give in to peer pressure and abandon the horses for wheels, no matter how skilled they are with horses and how much they seem to enjoy them.*

Over the years that I've practiced there has been little change in this. At junior horse shows, where competitors are eighteen years or under, the ratio of girls to boys is about thirty-five to one. The boys who stick with the horses are the good ones, the most talented. By the time young horsemen and horsewomen reach their twenties the ratio is nearly even.

Joe was one of those who stayed with the horses and, like other young men who continue in equine competition, he was good. He was offered a scholarship to study at a college that teaches horsemanship and riding, but instead he decided to go to a two-year business college and join his father in the family's thriving glass business. His now "aged" Ebony was still a good mount for a one-day show, but Joe wanted to compete in three-day events, which include dressage, stadium jumping, and cross-country jumping. This combination, known to its participants as "eventing," is the most grueling and competitive type of showing that a horse and rider can compete in. Joe, like many young men, liked nothing better.

At eighteen he began looking for the right horse and at nineteen he was still looking. When Joe's parents bought Ebony for him it was a perfect match of a young rider and his first horse. He had already ridden the horse for a year, and it was purchased from a trainer/dealer whom they could trust. Finding the right horse for a skilled rider in three-day eventing is a far more difficult task.

There are two ways to add horses to your ownership. One is by natural addition, breeding and raising your own. We shall discuss this in following chapters. The most usual way is by purchase. Let's look into the details of buying a horse for Joe Howard, a skilled rider, for the special purpose of three-day eventing.

How do you go about finding a satisfactory horse? How does anyone with better than average horsemanship and skill, expecting top competition, find the right horse?

Money is not the answer, although top horses are not cheap. Paying $30,000 for a horse is no guarantee that it is going to be any better suited to your needs and wants than a $3,000 horse. People are the main factor in your selection. You, of course, are the most important of these people since it is your money and your desires that must be satisfied. The other people are the professionals you consult and eventually deal with. These are the trainer, dealer, veterinarian, farrier, insurance agent, and perhaps even your attorney.

PROFESSIONALS

There are dealers and trainers who specialize in handling certain types of horses. Some buy, train, and sell; others act as agents for owners who wish to sell horses. Most fully trained and conditioned three-day-event horses are owned by people who are showing them, and they are not for sale except at extremely high prices. You will have to look around for a dealer whose horses are not yet conditioned to withstand the grueling pace of competition, even though they have the ability and strength to win. The buyer of such a horse knows both he and the horse have hours of training ahead to achieve peak condition.

ASK QUESTIONS

You will have to travel miles to try different horses, yet you might find just what you want in the next town. The right horse will feel right for you in personality and you for it, but there are still many questions. The age and history of the horse are important, and sometimes unobtainable. The reputation of the agent and/or the seller is very important. Horses that were "just shipped in" from 1,500 miles away, whose history cannot be confirmed, should be looked on with suspicion, as should those that have been for sale for years.

You yourself can look for old scars and conformation faults that can cause breakdowns. You can work the horse hard and listen to its breathing sounds to be sure it is not a "roarer," or broken-winded. You can also bring in your riding instructor or coach, or a friend who is also a skilled rider, to try the horse. You can have your friend take videotapes of you riding the horse and watch them over and over, not only to learn how sound the horse is and judge its appearance, style, and quality under saddle, but also how it and you appear as a team. Do you complement each other, showing enough "class" to catch the judge's eye?

EXTRAS

There are some things you can do that some might think unconventional, but with the kind of money you will have to pay for this horse you must consider everything. You can ask to load the horse or see it loaded, on a trailer and a van, to make sure it will not be a problem taking this horse to shows. You should take the horse over an outside course alone and with other horses. If at all possible, this should be tried during a rainstorm with wind and thunder! Then, too, you should take the horse out where there is traffic and commotion, things that must not bother any show horse. If it reacts badly to the aforementioned things it could make the horse worthless for the purpose for which it is bought.

After all this, and with a day or more to think it over, you should get a final agreement on price and terms including a provision that the horse pass a pre-purchase exam done by the veterinarian of your choice.

THE PRE-PURCHASE EXAM

The pre-purchase exam in this case will cover far more than the exam done on the young racehorse in Chapter XII, since this horse will be mature and is intended for one of the most stressful uses of a horse. It should include, but not be restricted to: radiographs of all four feet, and perhaps cannon, knee, and hock; a fiberoptic scope exam of the horse's respiratory tract; an opthalmoscopic exam of its eyes; an ultrasound exam of its tendons; a Coggins test with a repeat in thirty days; blood chemistry and a test for drugs such as painkillers, tranquilizers, and those that affect respiration and circulation. Sellers expect this today. It protects them as well as the buyer. In addition, any tests the veterinarian decides to do should be agreed to, including determining the age by teeth if the horse is not a tattooed Thoroughbred. Other tests, of course, such as stressing the horse and listening to its heart and lungs are routine in any pre-purchase exam.

Of passing interest, it should be mentioned that horses must have shoes removed in order to do radiographs of their feet. Even if the radiographs are not done, never buy a horse with pads on his feet until the pads are removed and the bare feet examined.

PAPERWORK

If you do buy the horse, and pay top dollar, you should have the horse insured as soon as it passes the pre-purchase exam and the sale becomes final. In addition you will probably want to update all immunizations and wormings. We can be sure that when you bring this horse home it won't be turned into the paddock with strange horses without care and pre-conditioning.

CHAPTER XV

APPALOOSA HORSE CLUB

BREEDING A MARE

*C**opper Penny** was well named according to her beautiful color, but in size she more resembled an old-fashioned silver dollar, being as large as any registered American Quarter Horse I have ever seen. Penny was purchased with a foal by her side at a Quarter Horse farm dispersal. It appeared that the foal would grow to be a duplicate of its beautiful mother. Included in the purchase package was an unseen individual, presumably a full brother or sister to the foal, since Penny was said to have been bred again to the sire and had been determined to be pregnant before the sale.*

The buyer was a young girl, intensely interested in Quarter Horses. Although not yet a teenager, she was interested not only in riding Penny, but also in continuing her use as a brood mare. The good type and excellent blood lines of the mare and the talent of the young owner seemed to make an excellent combination.

The mare was known to have a few bad habits, such as being headshy, but with her new mistress she soon got over this and was a perfect animal to work with and to ride.

The existing foal was weaned and the older he grew the more beautiful he became. Everyone familiar with the situation complimented the youngster on her purchase, telling her that "soon you'll have three of a kind." They always exclaimed about the two horses' beautiful solid burnished copper color.

The mare's due date was early May and here again Fortune smiled. A warm early spring meant that the mare could foal outside, or at least could be allowed to run outside with her new foal from birth on except in cold rain. Since most mares foal during the early morning hours, the mare was kept at night in a small grassy paddock, lighted so she could be watched, and during the daylight hours was turned out in a beautiful rolling pasture. With the good weather and Penny alone in the pasture the grass was as tall in early May as it might be some years in early June. Already the green grass was spotted with patches of wild daisies.

Cindy, the new owner, watched the mare carefully for waxing and other signs of imminent foaling. As she turned her out each morning she checked her particularly closely and as the school bus ground to a halt each afternoon Cindy jumped off and made a dash for the pasture to check Penny again. During the night, even though there was little sign the mare was ready to foal, Cindy still got up every three hours to go out and check. In between she slept little, kept awake by excitement and anticipation.

On May 7, a warm beautiful day, when the school bus stopped, Cindy couldn't see Penny in the near end of the pasture where she usually waited. Her heart pounded with excitement, yet she knew that in the morning Penny had shown no more sign of foaling than ever. Her parents and the veterinarian had assured her she would have plenty of warning and that Penny, like most mares, would foal during the quiet evening or early morning hours.

Cindy ran through the gate and raced across the field. As she reached the limestone ridge in the middle of the field, which blocked her view of the far corner, she saw Penny standing eating grass. Behind her, just clearing the tall grass and daisies, she could see the copper-colored head of a foal. Cindy couldn't see the entire foal over the big patch of daisies.

She was already enough of a horsewoman to know not to run up too quickly and frighten the foal. She called to Penny and, with her heart pounding from running and excitement, she walked slowly through the grass to see her third of "three of a kind." The daisies kept blocking her complete view until suddenly, despite her careful movement, Cindy could see the foal draw its front feet and legs together and, like a mature horse, raise first its forequarters, then its hindquarters. Strong, but still not quite steady, it stood with its copper-colored rump still seeming to be covered with daisies.

But to Cindy's horror, she realized they were not daisies. The foal was beautiful, although it was not a filly as she had hoped. But that didn't really matter, nothing mattered any more to Cindy. Her long-awaited, solid-copper-colored purebred American Quarter Horse had a perfect white spotted Appaloosa blanket!

Now in her final years of veterinary college, Cindy has probably had as many disappointments as all people who work with livestock. Still, nothing has ever bothered her quite as much as finding that day what most people would have been thrilled to find, a perfectly marked Appaloosa, when for nearly a year she had counted on a perfect solid-copper-colored American Quarter Horse.

The explanation of the Appaloosa foal born from a copper-colored Quarter Horse mare was easy to understand in one way and difficult in another. The farm Penny came from did have a well-bred Quarter Horse stallion, the sire of the first foal. This stallion had been bred to the mare. However, also on the farm was a Pony of America (POA) stallion, a breed some call miniature Appaloosas. How this tiny stallion managed to be in contact with Penny and, more difficult to explain, how he could ever have reached her to breed her was a puzzle. Some farms use pony stallions as teasers. This was not said to be the case, and no one ever admitted knowing how physical contact was made.

The perfectly marked Appaloosa grew and developed all the Appaloosa characteristics, not only in color but conformation and disposition. Some people breed mares for years trying to get a perfect Appaloosa. Others, as people teased Cindy, get them simply by having a mare foal in a field full of daisies.

Equine reproduction has always been, and still is, a fascinating subject to me. Until the last fifteen years little really scientific work was done on it compared to on other livestock. Even today, horsemen skilled in the fields of riding, racing, showing, draft-horse management, and polo, don't seem to be at all knowledgeable about the subject of getting a mare ready to breed, the mechanics of breeding, and the anatomy and physiology of the mare and stallion. Worse yet, among both professional and amateur horse people there are more "old wives' tales" and downright superstitions about equine reproduction than about any other field in animal husbandry.

PLANNING

Deciding to have your mare breed and foal is not a decision one should take lightly. Neither is it easy to accomplish, even when everything is done according to the best modern knowledge. On some stud farms a 45 percent foaling rate is all that can be attained. Yet dozens, probably hundreds, of foals born each year are a complete surprise to the mares' owners. "Wild" horses reproduce so fast that they are a national problem on western rangeland. Mother Nature seems to have her way in spite of human management and mismanagement.

Breeding a mare is a very serious step. One needs to consult with professionals once the decision is made. But first, do you really want the responsibility of foaling the mare out and caring for the foal later? Do you have time and facilities to do so? Just as important, is your mare really of good enough quality to raise a foal and is she of the right disposition to care for one?

THE STALLION

If your answer to the above is positive, to what stallion are you going to breed your mare? The female seems to be dominant in passing on most characteristics, as in most domestic animals: that is, some mares throw good offspring regardless of what they are bred to. However, one should never breed to "just any stallion." This is contrary to ancient belief that the male is most dominant, only recently proven false. There are the exceptions, the stallions that pass on certain good and bad characteristics regardless of what they are bred to. Then, too, stallions of certain blood lines "click" with mares of certain blood lines to throw great offspring. Far too many colts, however, are left intact (not castrated) for sentimental reasons or, in fact, simply through neglect. Never breed to one of these, unless its offspring are known to be of desirable quality.

Normally the owner of a purebred mare will want to breed her to a stallion of the same breed. Sometimes, however, this is not the case, such as deliberately crossing a Thoroughbred stallion with a Percheron mare to produce a classic hunter. Another rule of thumb is to study the offspring of stallions and look for characteristics most lacking in your mare.

Breeding

Few horse owners ever own, or want to own, a stallion for breeding purposes, but if you are going to breed a mare you need to know what actually happens at breeding time. And before you attempt to raise and use your own breeding stallion you should spend some time observing and working in the breeding shed of one of the large breeding farms. As with any other art, stallion handling and breeding is best learned by doing.

The mechanics of breeding vary all the way from "pasture breeding," almost as in the wild, to artificial insemination and even embryo transplant. So-called "hand breeding," natural service under controlled conditions, can be a dangerous thing for the mare, for the stallion, and for the people trying to control it. Experienced stud managers sometimes make it look easy, but even under the best-controlled conditions accidents do happen.

NATURAL DANGERS

One might then wonder that if it is so difficult, how does conception ever occur in the wild, or in the pasture, and since it does, why not just turn the mare and the stallion together and let nature take its course? Nature taking its course can be dangerous. Stallions used for pasture breeding learn from experience which mares they can actually cover, and when it is safe. The experience can mean a broken leg for the stallion or, just as bad, a debilitating, crippling wound to his penis or testicles. Watching a wise, experienced stallion at pasture you will note he never rushes into anything, and usually "quarters" up to the mare, approaching at a 45° angle. In the wild there are forty times as many stallions born as needed, so that if thirty-nine get disabled there could still be one survivor to cover forty mares and live.

Conversely, mean stallions can hurt mares, particularly young or old mares, by crowding them into corners and rupturing a vagina or rectum, and/or severe biting on the neck or even the buttocks and genitalia.

HANDLING A STALLION

Should you decide to raise a colt that may be a stallion prospect, or for that matter any foal, start to halter-break him from day one and let him know you are the boss. One does not have to be cruel to be firm. In fact, a horse broken by cruel treatment will either be a shivering coward or a sly enemy waiting for the day the cruel human lets down his guard.

There are plenty of small farms with stallions properly trained from birth to respect voice commands. A few mares are bred each year with a minimum of commotion and fuss. It can be done, but one must never forget that every once in a while the excitement raised by hormones at breeding time can override the "manners" of the best-trained mare or stallion.

Common-sense precautions, such as using unshod stallions, never attempting to breed a mare that is shod behind, always bandaging tails, and being extremely clean, are a few basic rules you should follow whether your stallion is breeding one mare a year or forty.

HAND-BREEDING

Actual hand-breeding a mare should follow an unchanging routine. The mare is brought to the breeding area, inside or outside, which should be a clean open space with good plank fencing or walls, with no machinery, equipment, or debris around that animals can become tangled in. She should be wearing a good strong nylon halter and be controlled by a chain shank. Bandage her tail from root to tip and wash her buttock and genital area with soap or a mild detergent disinfectant. Wipe her dry, using disposable paper towels, working from the vulva outward so that the vulva and immediate surrounding area are as clean as possible. She may then be rinsed with clean warm water and again wiped dry from the vulva outward.

Then place the mare behind a *teasing wall*, a solid plank wall four feet high and eight to ten feet long. Bring in the stallion, controlled by a chain lead shank with a long lead (two shanks with two handlers or a short whip may be used in some cases), and lead him slowly along the teasing wall to determine if the mare is in standing heat. The stallion is allowed to nuzzle the mare, but not to get up on the fence. If the mare squats, "winks" her vulvar lips and seems receptive, the stallion should be backed away by voice command, with his rear quarters in a corner. Wash his now "dropped" penis, first with a mild disinfectant or soap and water, and then rinse.

The mare is then led from behind the wall to face it or some other solid wall where she can't move forward. A tail rope (see diagram) should then be applied and her tail pulled up to her neck. If a tail rope is not used one person may have to grab her tail as the stallion mounts. It may be necessary for two people to hold the mare, one on either side, using two shanks, or a nose clamp and a shank. The use of chain twitches is common, but if misused they can make the nose numb and thus become ineffective just when they are needed most. *Breeding hopples* (see diagram) are also commonly used, but cause problems on some mares. Both hopples and tail ropes should have a safety quick release to be used if a stallion catches a foot or leg in them.

In the case of valuable stallions, the actual breeding stallion is not used to tease, and the mare is "test-jumped" by a teaser stallion to be sure she will stand, the teaser being pulled away before he enters.

When the breeding stallion is brought up to the mare he is not allowed to come in at a run, to charge on his rear legs, or to frighten the mare. It is at this stage that the stallion is most difficult to control unless extremely well mannered. He should not be allowed to mount until his penis is nearly erect, but neither should he be held back until his glans penis is so large he cannot enter the mare. As the stallion mounts, one person may need to pull the mare's tail aside if it is not tied forward, and direct the stallion's penis against the mare's vulva. Some stallions don't need this, however—and some resent it so badly that they will back off.

Once entered, the stallion should be left to himself, though some handlers keep one hand on the stallion's foreleg, seeming to calm him. If a stallion is a biter he should either be muzzled or the mare's neck and withers covered with a heavy blanket or pad. Assuming all is well, the stallion will show that he is ejaculating by "flagging," a rhythmic motion of his tail usually easily recognized when one is used to a particular stallion.

As the stallion dismounts a quantity of semen will usually run out of the mare's vulva. This may be caught in a clean paper cup and checked under a microscope for sperm motility and proof of ejaculation.

Immediately upon withdrawal the stallion is again backed into his corner, hopefully by voice command, where his penis is washed and rinsed. The use of strong disinfectant either as a pre-wash or post-wash is discouraged because this may destroy the normal bacterial flora of the stallion's sheath, and allow harmful bacteria or molds to take over. Contrary to the washing of the mare, on the stallion simply washing away actual dirt is more important than trying to be sterile.

ARTIFICIAL INSEMINATION

Artificial insemination is almost always used on Standardbreds, sometimes on Arabians and other breeds, but is not permitted on registered Thoroughbreds. If you take a mare to a farm where artificial insemination is used, semen is obtained by using a "decoy" mare and artificial vagina. This method is similar to natural service except that the stallion's penis is directed into an artificial vagina, a device with a rubber liner surrounded by an area of air and warm water to simulate the sensation of the mare's vagina. One drawing of semen treated with antibiotics can be used to inseminate a

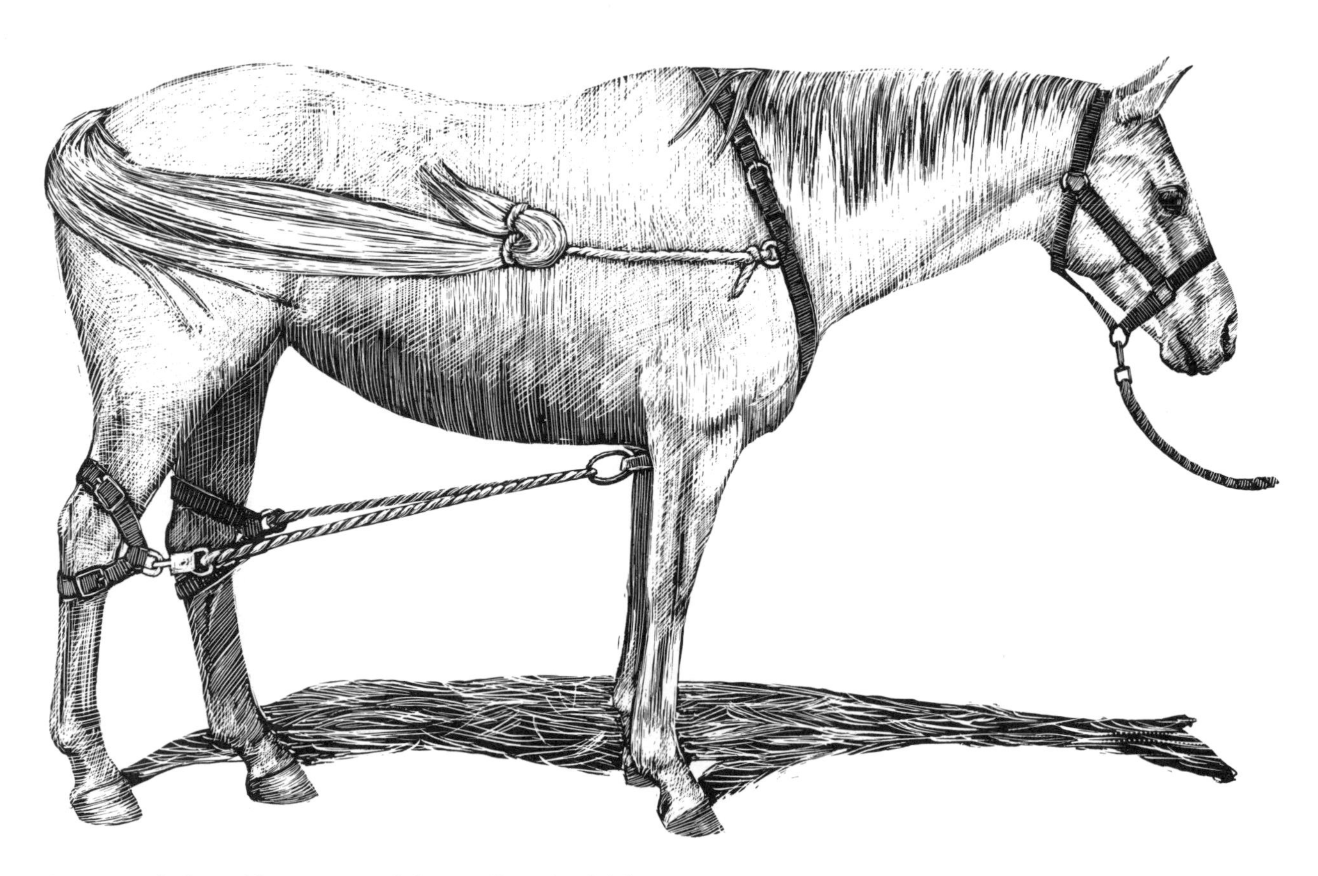

Mare with tail rope and breeding hobbles

dozen or more mares.

Stallions' semen may be frozen and stored, as is semen from other animals. However, determining exact heat and best breeding time in mares is so difficult that commercial use of frozen semen in artificial breeding units is rare.

If you wish to breed a mare to a particular stallion without transporting either mare or stallion your veterinarian can obtain fresh semen storage and shipping containers from companies supplying such equipment, and take care of it for you. It takes some advance planning and entails extra expense. Still, if your breed association honors this type of breeding, or if you are interested in a special-function foal, such as a hunter with no registration papers, this may be the way to get your mare bred. As a matter of disease control, of course, artificial insemination is way ahead of natural service.

It is often asked why A.I. is not permitted in Thoroughbreds. The answer has many facets, but the one I believe is most understandable is that Thoroughbred racing is a sport with very strict rules. The long history of the breed has always emphasized that a Thoroughbred should be able to reproduce with no outside help from artificial means. To perpetuate blood lines where certain individuals can only reproduce by artificial means, as has happened in some canine breeds, could make the Thoroughbred horse a creation of man in years to come, instead of a beautiful animal created by nature. I respect this thinking and feel the sport and excitement of Thoroughbred racing is the better for it.

Before sending a mare to be bred you should have her examined by a veterinarian familiar with equine reproduction. Before you do that, however, you should know a little about the anatomy and physiology of the mare in order to better understand what is entailed in breeding and communicate better with your veterinarian and others you will be working with.

The Mare's Anatomy and Physiology

The mare's genital tract consists of the vulva, an external opening just below the anus. The vulvar lips enclose the clitoris, a sensitive walnut-sized mound that is seen when the mare urinates or "winks" when teasing. Above the clitoris is the urethral opening from the bladder, and above that the sleeve-like vagina. The vagina extends forward to the brim of the pelvis and leads to the mouth of the uterus, or *cervix*. In the nonestrus (not in heat) mare the mouth of the cervix, viewed through a speculum, is almost at dead center of the end of the vagina and resembles a half-open rosebud. When the mare is in heat and ready to breed, the cervix lies on the floor of the vagina, is open up to three quarters of an inch, and appears to have clear shiny mucous flowing from it.

The mare's uterus, like that of most domestic animals, is *bicornual*, that is, it has two "horns" extending forward from the short body of the uterus. In the open mare these horns are one to one and a half inches in diameter and eight to ten inches long. From the tip of each horn, suspended by the broad ligament, a sheet of tissue, that also suspends the uterine horn itself, extends the serpentine fallopian tube or oviduct, and an ovary. Mare ovaries are bean-shaped and vary from one by two inches to two by three inches, larger in the older mare.

ESTRUS CYCLE

Depending on breed, climate and time of year, and influenced by the number of hours of light per day, mares show estrus or "heat" every twenty-one days. The heat period in the mare lasts from six to eighteen days, but usually the mare is really receptive to the stallion only on one or two of these days, just prior to ovulation when the egg is released from the follicle. Some mares only cycle a few times during late spring and early summer, with June 21, the longest day of the year in the Northern Hemisphere, the peak of breeding activity. Others, the Thoroughbred in particular, may cycle year-round. If left to nature most mares in a wild state in North America would foal in late May and be pregnant again ten to thirty days later.

This is one of the most remarkable things about equine reproduction—that most mares will have a so-called foal heat an average of nine days after foaling and, if bred at that time, may conceive and carry a foal to term. During this nine-day period the uterus shrinks from a sack large and strong enough to hold a 100-pound foal, placental membranes (afterbirth), and fluid weighing about another 100 pounds, to an organ with each horn no larger around than a loaf of French bread and about ten to twelve inches long.

OVULATION

At the time of the mare's birth each ovary contains the potential to produce hundreds of ovum, or eggs. During her second or third summer her first heat occurs, in which a so-called follicle, a fluid-filled half-bubble containing an egg, enlarges until it may be as big as the ovary itself. Ovulation occurs when the follicle ruptures, releasing the tiny, almost microscopic, egg, which is then picked up by the fimbrionated flower-like mouth of the oviduct. Two to six days prior to rupture of the follicle the mare will show physical signs of heat, which will be described later.

Immediately after ovulation, the ovary will show a pit-like depression (where the follicle was) that rapidly fills with blood. This can be detected by rectal exam. This blood-

MARE'S REPRODUCTIVE ORGANS

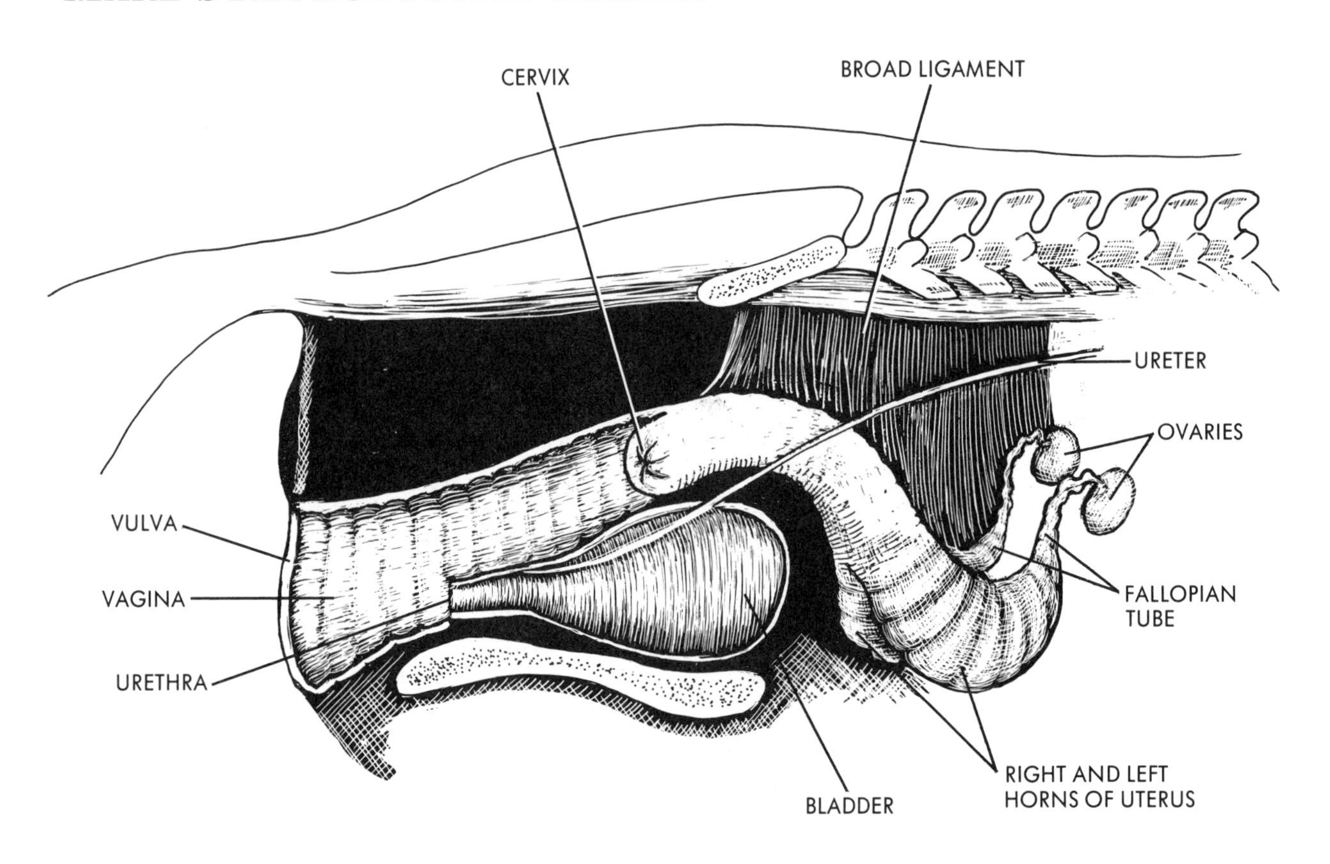

clot is called the *corpus hemorrhagicum*. At this time, a few hours after ovulation, the mare will no longer accept the stallion.

As the blood clot hardens it changes to a yellow-orange mass called the "yellow body" or *corpus luteum*. The *corpus luteum* grows for about a week, stays the same for a few more days, and if the mare has not conceived it starts to shrink (regress). During its active time, it produces a hormone, progesterone, which keeps follicles from enlarging and thus the mare from showing heat. If the mare does not conceive about the time the *corpus luteum* starts to shrink, another follicle starts to enlarge and the cycle starts all over again.

If the mare has conceived, however, the fetal sac of the embryo secretes a hormone that causes the *corpus luteum* to continue producing progesterone. This prevents another cycle by keeping other follicles from enlarging.

CONCEPTION

If the mare has been bred in the days previous to ovulation and/or had sperm from the stallion introduced into her genital tract, they will have migrated through the uterus up into the oviduct. These sperm will surround the egg, and one will enter and fertilize it. The fertilized egg takes about six days to pass down the oviduct to the uterus. When it reaches the uterus the egg, now an early embryo, will remain there, nourishing itself on uterine fluid. In about thirty-five days, as it grows and forms its own fetal membrane sac, it becomes attached to the uterine wall. By twelve to fourteen days after fertilization this sac and embryo may be seen and, in fact, photographed, using ultrasound equipment.

By seventeen days after fertilization (or conception) a veterinarian skilled in equine reproduction can tell by rectal palpation, with 90 percent accuracy, if the mare is pregnant, by the tone of the uterus and the shape and texture of the cervix. By twenty-three to twenty-four days after conception, the veterinarian can often feel the tiny fluid-filled fetal sac by rectal exam.

During the first pregnancy check, whether by palpation or using ultrasound, the veterinarian tries to be sure that the mare is pregnant with only one embryo and not carrying twins. Twin pregnancies in the mare most often end in abortion and death of both twins unless one fetus reabsorbs prior to thirty-five days. A veterinarian finding a twin pregnancy in your mare will discuss the situation with you and attempt to either cause one embryo to reabsorb or to abort both, unless requested not to.

If conception has not taken place the egg will stay in the oviduct and eventually deteriorate. In 5 to 10 percent of cases the fertilized egg or embryo does not "take" in the uterus and disappears, termed by some as "reabsorption." There are many reasons given for these early embryonic deaths, one being that they are nature's way of eliminating imperfect embryos and avoiding the birth of imperfect foals and/or twins.

The foregoing is an oversimplification of a very complex process which, in the mare, has so many variables that a researcher spending a lifetime studying the reproduction cycle of the mare could never unravel nor attempt to explain them all. As the owner of a mare you wish to breed you need to know at least this much to understand what to do to bring your mare to the stallion at the correct time.

To be sure, you will find individual horsemen who know all the answers, but this is only because their limited experience has not exposed them to all the questions.

CYCLING BEHAVIOR

In addition to the above, you should know by the mare's actions which stage of her reproduction cycle she is in. During midwinter most mares give little evidence of cycling. In late February a cycling mare in early estrus will react to a stallion or a strange gelding brought along a fence of her yard or stall by first approaching him, nosing him, and then whirling and kicking at him. A few days later she may not whirl, but put her rear quarters toward him and rub him. If he makes any attempt to return the affection by reaching over the fence and nuzzling or nipping she will squeal, perhaps kick, and run off.

Two days later, if this is a true cycle and not an early-season prolonged "false" heat, she will rub against the fence when the stallion is presented, squat, jet urine, raise her tail and "wink," turning the edges of the vulva out as though to invite the entrance of the stallion's penis.

WINTER HEATS

In winter some mares reach any of these stages and will then level off, some accepting the stallion day after day for up to two weeks, but not conceiving. Examination of such mares by a veterinarian will reveal that they have a follicle that either doesn't change, or regresses (shrinks away) without ever ovulating. The examination of the vagina and cervix, either visually through a lighted speculum, or by palpation with a gloved hand, is another method by which the veterinarian can tell if the mare is ready to breed. In a mare that is ready to breed the cervix should be open and flowing and the vagina moist and slippery.

If the first heat is a false heat the next is more apt to be normal, although from late winter into early spring most heats tend to be prolonged and the conception rate low. One might think that the use of hormones, as used in other species, would increase the chances of ovulation and pregnancy, but their use in most cases is disappointing.

The use of artificial light does seem to start normal heats in mares earlier in the year. Starting about December 15, if a 200-watt bulb is put on in a mare's stall in early morning and left on for sixteen hours, a healthy mare will begin to have normal heat cycles in about six to eight weeks. Once she is bred and pregnant, the light must be continued until the mare goes out in late April or May, or what was diagnosed as a thirty-day pregnancy in March may no longer exist in May.

Even the stallion is not as fertile in winter. In February sperm may live only for minutes out of the stallion's body. In June they will remain alive in the mare's fallopian tubes six days or more.

Unless you are trying to breed a mare of one of the racing breeds, however, and particularly if you are breeding a mare to produce an ordinary riding horse, you are far better off to aim for a May breeding, which will give you an April foaling at the earliest. In this way your mare will foal during warm weather and you don't have the problems associated with cold and wet weather.

Racehorse breeds use January 1 as the birth date for all foals, even those born in December. The gestation period of a mare is about 335 days (eleven months). To give foals a competitive edge, emphasis is put on breeding racehorses as soon after February 15 as possible so that they will be born soon after January 1. Other breeds do not usually aim at breeding this early in the year.

Preparing the Mare

Now that we have discussed the basics of mare reproductive physiology we can more easily discuss how you prepare a mare for breeding.

As soon as you decide you might breed a mare, start trying to detect possible heats and note them on a calendar. When you have only mares and no gelding or stallions around it is almost impossible to detect heats on some mares. If you are lucky though, a lone mare will show a slight change in disposition, such as being cranky or more affectionate than usual. Many mares "wink" when they urinate, but if a mare doesn't usually do this and you note that she now does, put down the date and see if she repeats it three weeks later.

On rare occasions where two mares are together one may "ride" or mount the other as cattle do. In spring, or in fact at any time of the year, a gelding may mount a mare in heat and actually enter her. This, though a good indicator of heat, is also a way of spreading serious disease. Bringing in a strange gelding or putting a mare in close proximity to a stallion will often cause her to show heat when she has appeared dormant around her usual stable mates. Some mares will actually go through the squat, urinate, "wink" routine known as teasing even when alone.

FALL PRE-BREEDING EXAM

Once you have a rough idea of a mare's cycle have the mare examined in September or October by a veterinarian knowledgeable and skilled in equine reproduction. If the initial exam is made later than early fall the mare may be in winter anestrus, that is, the ovaries may be dormant, and the examiner has no way of knowing if the reproductive tract is functioning normally.

It is best to schedule the September or October exam during the time you expect the mare to be in heat. If the veterinarian suspects infection in the mare he or she may decide to culture her uterus, which is of practical value only during estrus. Culturing is done by passing a sterile cotton swab into the uterus through the cervix. Usually a speculum and special instruments are used, although some veterinarians use a double-glove technique with good results.

The swab is then applied to special media that will grow bacteria present on the swab. These bacteria are identified and sensitivity tests are made to determine which antibiotics, if any, will work to eliminate the bacteria.

Another common test made to determine the fertility of a mare is a biopsy, that is, to remove a piece of inner lining of the uterus and examine it microscopically. This is usually done on older mares with poor breeding history, or if the veterinarian noted suspicious lesions on rectal exam.

One other common procedure on older mares with poor conformation is to suture the upper part of the vulva opening to prevent "wind sucking" and help clear up chronic vaginal infection. If you own a mare that was once raced you may find she was already sutured. A common name for this technique is the Caslick operation.

There are other tests and procedures, but the usual fall pre-breeding exam involves only a rectal exam to palpate the uterus and ovaries and a speculum exam to view the cervix.

ARRANGING FOR A STALLION

Hopefully your veterinarian will find the mare to be a good candidate for breeding. The next step will be to decide on the stallion you want to use and to make contact with the owner or, in the case of racing stallions, the agent. Fall is not too early to find out the health requirements of the owner or agent, such as tests required, cultures, the breeding fee, and such arrangements as whether you bring the mare in on breeding day and right home again. Some stud farms prefer to keep the mare until she is checked pregnant. The stallion owner may also require that the mare be brought in three weeks ahead of time to have his own veterinarian examine her and culture her.

Some stud owners require payment of the breeding fee in advance, others when the mare is checked pregnant, some the following September 1, and some only after a foal is born and standing. Regardless of when you pay the fee, it generally covers a "standing foal," with refund or another breeding if this is not the result. Besides the stud fee there may be certain fees for services at the time of breeding, such as examination, insemination, and boarding. Be sure to find all this out before you sign a contract.

If your mare goes to the breeding farm three weeks or more before breeding you may only have to have your own veterinarian check her once or twice prior to breeding. You will probably need a Coggins test for EIA and whatever other tests or immunizations the stallion owner requires.

If you have to take the mare to the breeding farm only on the day she is ready to breed and you have no way of "teasing" her you may need to have your veterinarian come once a week beginning in February to examine for possible heat. Then, on the predicted first day of the heat period when you wish to breed her, your veterinarian will have to examine the mare and repeat every forty-eight hours until ovulation can be predicted by palpation of the ovaries and checking the condition of the cervix. Forty-eight hours later your veterinarian will have to recheck your mare to be sure she has ovulated. If she has not she should be bred again ("covered" as in natural service or "inseminated" as in artificial insemination).

Whether it is required by the breeding farm or not, your veterinarian will recommend certain immunizations prior to and after breeding, such as worming, pregnancy rechecks, and perhaps a special vitamin-mineral mix to be fed during pregnancy.

The mare should be checked by ultrasound between fourteen and nineteen days or by rectal exam between seventeen and nineteen days, then again at thirty days.

Once your mare is home and checked to be three months pregnant you should begin planning for foaling, which should occur about 335 days from breeding, but can happen normally any time from 300 to 365 days later.

CHAPTER XVI

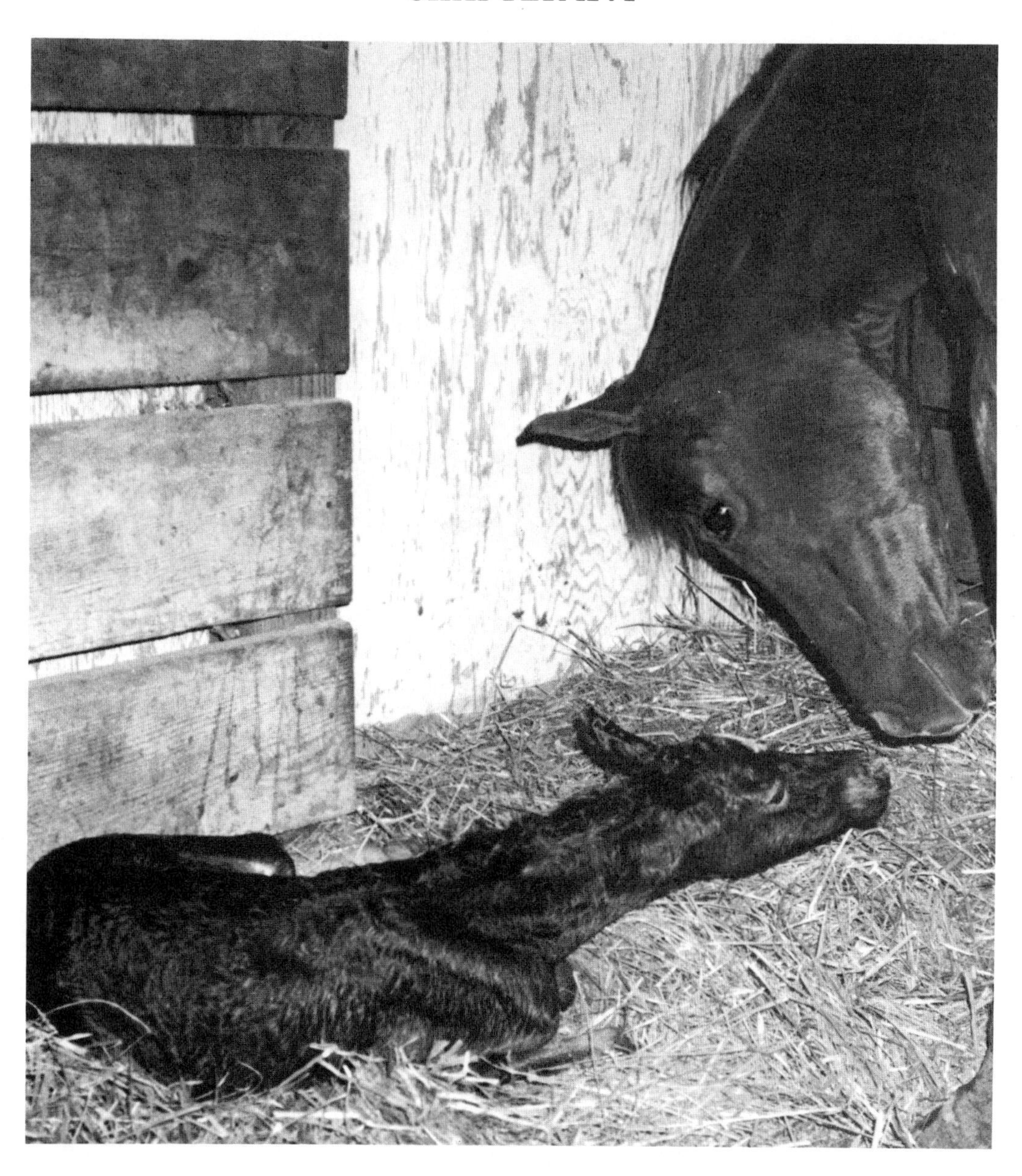

INTERNATIONAL ARABIAN HORSE ASSOCIATION

FOALING

*A **few years ago** I was to meet a client at a farm outside of my usual practice area to examine a young horse he was considering buying. The appointment was for one o'clock but, in an unusual occurrence, my technician and I arrived at the farm about twenty minutes early. Since all the farm employees were at lunch and it was a sunny day in late June, I pulled the car under a shady tree near a grassy paddock with the intention of looking around while we waited.*

As we got out of the car we noticed an obviously aged and pregnant Standard-bred mare sharing the shade on the other side of the fence. She stood motionless on three legs with one rear leg resting, her head down, with only an occasional switch of her tail at an invisible fly to let us know she was not a statue. My assistant, Claudia, recognized her as a mare she knew from having worked on this farm in years past, and spoke to her.

At first the mare seemed to pay no attention to the soft "Hello, Mary T." But then, in slow motion, she turned her head and looked toward us. Claudia remarked, "Look, she's sweating, and she's waxing."

As she said that a fine stream of milk no thicker than a spider's web jetted from one of the mare's teats. The other teat and the inside of the mare's legs were covered with honey-like "wax." This is the first colostrum, which hardens as it leaks from the mare. At the same instant Mary T.'s huge belly seemed to jump as if she had a case of hiccups.

Milk then came from both teats, four streams spraying in the sunshine as she started to move slowly to the center of the paddock. Next Mary T. turned her head back and looked at her belly as a horse with colic might, but seconds later she was eating grass as though she had not a pain in her body nor a concern in the world.

Five minutes of unconcerned grazing ended with another hiccup-like spasm and renewed spraying from both teats. Mary T. began to walk about the perimeter of the paddock, then in smaller circles, and once threw herself down, only to jump up with a groan and walk in smaller circles.

As she moved away from us her raised tail revealed a pale pink bulge from her vulva lips. Seconds later she was down again, gave two spasmodic strains and, with a grunt, was back on her feet. This time the bulge was bigger, but instead of walking she started to eat grass again.

In a moment she went down again, then was back up and straining. Suddenly there was the sound of fluid, like a pail of water being dumped. It happened so fast that the sound seemed to reach me quicker than the sight.

By now the sweat was literally pouring off the mare and milk jetted from her teats. Again she turned and, I thought, started to graze. But this time she was licking the ground where her "water" had landed. Now we could see two tiny feet in a yellow plastic-like covering poking out under the broken pink sac. She licked for a moment, then gave a half whirl and went down, over on her side, with great grunts. We could now see the foal's nose — two more grunts and the eyes, and then the poll and ears. Again she rested for a second, gave another grunt, and the long neck and legs, nearly to the elbow, were out. The foal shook its head as though to clear itself of the shroud of membrane.

There was no more resting now ... grunt, push, grunt, push, and the shoulders were through. The long body seemed to slide out on its own and then it was stopped by the hips. There were three great pushes and grunts and then a sigh from the mare as the hips came out and all that was still in the mare were the legs from the hocks down. Mary T. lay her head down and rested for a few seconds, seemingly exhausted. Then in a minute she raised her head again and we were aware she was straining slowly, easily. Blood and fluid poured out, and as the foal shook its head and kicked, the navel cord snapped but little more blood came. Mary T. rolled up farther in a more usual lying position, reached back, and nuzzled her foal.

Five minutes later the farm manager arrived and Mary T. jumped to her feet as the placenta dropped to the ground. "What are you gawking at Doc — never see a foal born before?"

"How did you guess, Ed," I answered, and I knew he wouldn't believe me. After forty years of practice and a lifetime spent

with horses, this was the first mare I'd ever seen foal as she would in nature, completely unassisted and without me and sometimes half a dozen other people trying to help.

In the last chapter I said equine reproduction was the most fascinating part of veterinary practice for me. Foaling is the most exciting. No matter how well you are prepared, and no matter how long you are in the business, foaling is exciting and also worrisome. One can never learn enough to meet every problem and solve it with success. Still, there are certain rules and procedures that a mare owner should follow. Of course, most mares foal unassisted, and forty-nine times out of fifty you will get a healthy live foal from every mare you breed.

HEALTH CARE

Let's start where the last chapter left off. You have your mare home after a thirty-day pregnancy check. Your veterinarian will recommend certain immunizations, the most important being rhinopneumonitis prevention and worming. You may be told to feed a vitamin-mineral mix in addition to the regular diet. If the mare was bred early in the season (March or before) you might ask your veterinarian to do a sixty-day pregnancy recheck, since it would still not be too late to breed her again. Otherwise have a check made at ninety days and then again in late September or October just to be sure. If she is open you can get ready to breed her in the spring.

The October check is a good time for a rhino booster, and if your mare has not had flu or tetanus boosters in the last year they can be given. These should be repeated at two to four weeks before foaling, plus a rhino booster given every sixty days from October on.

HOUSING

By October, if you do not have proper foaling facilities, you should start building them before winter sets in. If your mare is due prior to May you should have available a box stall 14 by 14 feet or larger, 16 by 16 being better, in a tight but well ventilated, dry barn. You may need a safe place to rig up a heat lamp for the first few hours of the foal's life if it is due in March or earlier.

Earth floors of gravel, stone dust, clay, or packed sand are preferable to wood, concrete, or blacktop in foaling stalls. These materials may be dug out and replaced every year or two, depending on the amount of use the stall receives. Whole straw bedding is a must for foaling stalls. Sawdust, shavings, peanut hulls, or chopped bedding of other materials are not satisfactory and should not be used once the mare is close to foaling. Clean whole straw is difficult to obtain. You may need to start looking for it months ahead.

FOOD AND WATER

During pregnancy the mare should get plenty of fresh air and exercise, with good-quality hay and more grain than an open non-working mare receives. Stay away from high-protein feeds, particularly late in pregnancy when overgraining of a ration containing lots of corn and high protein can cause udder edema, or "caking." As mentioned in Chapter III, using so-called complete rations (at one and a half pounds daily per hundredweight of animal) for pregnant mares, with restricted hay, will avoid "hay belly'' and leave room for the foal to develop, while keeping the mare in good condition.

A bran mash (See Chapter III) fed once a week is excellent for pregnant mares. Water should be available at all times, warmed a bit in the coldest weather when horses tend to become dehydrated more than in hot weather. This dehydration occurs either because a horse's water freezes, or because it is too cold and hurts their teeth.

SUPPLIES

By December it is not too early to start laying in the things needed for foaling. A list follows.

1. 12-quart plastic bucket, new and clean, and a pint or quart plastic or stainless steel cup
2. Tail wraps and muslin bandage that can be used for tail wraps
3. Cake of Ivory soap
4. A pint or more of a good disinfectant such as Nolvosan or Betadine
5. A few plastic shoulder-length obstetrical sleeves (which are actually plastic gloves with long sleeves attached)
6. An electric lantern or good flashlight
7. A heat lamp
8. Large Turkish towel
9. Paper towels
10. Non-sterile cotton, one roll
11. Four ounces regular-strength tincture of iodine
12. Two Fleet enemas, adult size, or foal enema kit
13. A large water-resistant paper bag to put placenta in
14. A barn telephone so you can be in touch with your veterinarian if you have a problem
15. Anything else your veterinarian suggests

Optional supplies:

- A pair of obstetrical chains and handles or a set of nylon pull straps and handle
- Oxytocin, 10cc (to use under veterinary advice if a mare does not clean)
- Syringe and needles
- Antibiotic such as penicillin/streptomycin
- Small emergency oxygen tank
- Surgical scissors
- Two hemostats (artery clamps)
- Barn scale
- A supply of obstetrical lubricant

These supplies should be stored in a clean, dust-free cabinet and not used until foaling. The oxytocin and the antibiotic should be refrigerated.

ONE MONTH BEFORE

About a month prior to the due date, which, as we mentioned, depending on breed averages about 335 days or eleven months from breeding, call your veterinarian and remind him or her that you have a mare due and would like her given a tetanus toxoid booster and, if the veterinarian feels it necessary, a flu booster. If at that time there is any doubt in your mind or the veterinarian's that the mare is still pregnant, have her rechecked. At this visit also ask your veterinarian if your mare had a Caslick operation or has been sutured. If so, she can be opened at this time, or an old mare with poor conformation can be left for you to open at foaling. If you must open her yourself be sure you have a pair of sharp surgical scissors to do the job when the time comes.

If a mare only nine and a half or ten months along suddenly starts to ''bag,'' that is, her udder suddenly enlarges, call your veterinarian sooner.

In mid-winter mares seem to run a few days over the 335 days. In spring when the mares due are outdoors exercising they foal closer to the 335 days.

When your mare starts to bag, or show rapid udder development, watch her more closely. If bagging seems excessive, however, forming hard edema that pits when you press your finger into it, it may be good to call your veterinarian. The mare may need to have her grain cut back.

There are devices such as closed-circuit television and alarms that go off when a mare rolls over on her side in the final stages of labor. If such would ease your mind they are worth the money. Still, most mares will simply be watched as they have been in the past.

ONE DAY BEFORE

The normal mare will give twelve to twenty-four hours' notice of impending foaling by "waxing." Waxing simply indicates that milk is forming in the udder and is starting to leak out. It usually looks like honey or beeswax and appears at the teat ends. It may be milky-white in appearance and literally stream from the mare's teat openings. By the way, a mare has two teats, like a goat, but has four quarters, like a cow. Thus a front and rear quarter have separate openings in each teat.

When milk streams from the udder for more than twelve hours, and foaling does not occur, it might be wise to collect as much as you can and freeze it. This is the valuable colostrum, the milk that imparts immunity to the newborn foal in compounds similar to gamma globulin. Don't remove milk manually from the mare, however, since it will encourage more let-down and eventual loss of colostrum. Most veterinarians will appreciate your call when your mare starts to wax rather than an emergency call should you have a problem.

Inducing. If a mare leaks clear milk for more than twenty-four hours you should call your veterinarian and discuss the situation. Parturition (foaling) may be induced by oxytocin injection, but most veterinarians will not do this unless the mare is "coming into her milk." The leaking mare obviously is, and therefore is a good candidate for inducing. The decision should be made by you and your veterinarian after weighing all the facts, advantages, and possible hazards.

Once started, a mare induced with oxytocin (posterior pituitary extract, purified oxytocin principle, known as P.O.P.) will foal and clean within forty minutes to an hour. I have seen mares foal unassisted in twenty minutes from the first injection. I have also had mares I pronounced "not ready" for inducing after a vaginal exam foal an hour later with no injection. A veterinarian will never induce a mare without complete examination and knowledge of all the facts.

PLAY IT BY EAR

On occasion mares will foal without waxing, or even bagging. In fact, after such a foaling some will appear to have absolutely no milk, and three hours later are streaming milk. As in breeding mares, the longer one lives and the more mares one sees foal, the more variations are noted. All one can do is to "play it by ear," gain knowledge from experience and not be ashamed to call the veterinarian for advice and assistance if needed. Most veterinarians would rather have you call and ask a question about something normal and easily explained than have you wait until you are in serious trouble.

Don't be discouraged. Your mare will probably run over the 335 days, wax, and then foal unaided in the ten minutes you take to go to the house for coffee after you have stayed up all night watching her doze in her stall. If she bags, waxes, and then stops waxing it may be normal, too, but better phone your veterinarian and have her checked.

A night man in the foaling barn on a stud farm had difficulty reading, yet he could predict twenty-four hours ahead, almost to the hour, when a mare was going to foal.

Don't expect your veterinarian to be able to say when a mare is going to foal. He or she has not seen your mare each night for the past hundred nights, and can't note the subtle little differences in each mare as can a person with years of experience like a night man in a foaling barn.

You, however, may be enough of an observer to note the differences in your own mare. The thing we all can see is the waxing, the enlargement of the vulva, and the relaxation of the pelvic ligaments. As the owner of a mare you may notice a dreamy "not tonight" look in the mare's eyes, which often does mean she is going to foal. Other signs are sweating, pacing, getting up and down, finally actual straining (pushing), and the appearance of the first pink bubble of placenta.

One might say that symptoms of foaling are similar to colic. The difference is that the foaling mare will still eat and the colicky mare will not.

Foaling

At the first indication the mare is going to foal, and hopefully before the placenta shows, be sure to bandage her tail. If you have time and don't seem to upset the mare by being in the stall, wash her teats and udder. If her legs and buttocks are dirty wash them, too, with soap and water or a mild disinfectant such as Nolvosan or Betadine. Then put more clean straw bedding in her stall (never use sawdust in a foaling stall), get out of the stall and stay as far away and as quiet as you can while still able to know what is going on.

FOALING PROBLEMS

Before you read further, go back and read the first part of this chapter. I have tried to describe normal foaling as it occurs most of the time.

A rule you should always follow, whether you are experienced with foaling mares or not, is that if you have any doubt, even if you have a "gut feeling" that your mare is in trouble, phone your veterinarian, explain the situation and let the veterinarian decide whether to come or wait, or to tell you how to correct your problem.

The normal presentation for a foal is two front feet appearing, followed by the nose. I will describe some malpresentations in which foals may be found at foaling. Some of these you can easily correct. But with others, as soon as you realize what you have you should call your veterinarian for help, and only try to correct them yourself if you can't get help. Unless your veterinarian is very nearby, most of these will result in a dead foal, but at least your veterinarian can get them out without hurting the mare. We shall describe some malpresentations that even an experienced veterinarian finds difficult, if not impossible, to correct short of a Cesarean. We describe these so you might better appreciate what your veterinarian is up against.

Never attempt to reach into a mare unless you have:

1. Bandaged her tail;

2. Washed her genital area and buttocks with a mild disinfectant such as Nolvosan or Betadin;

3. Cleaned your fingernails, scrubbed your hands and arms, and put on a clean plastic sleeve;

4. Lubricated the sleeve with soap or lubricant;

5. If at all possible have someone to hold the mare for you.

ONE FOOT PRESENTING

If instead of two feet appearing you see only one, after following the above five rules, reach in and feel whether the other foot is right there where you can get it. If it is and you can't pull it out to be even with the first one, get the mare on her feet, and push the

POSSIBLE PRESENTATIONS OF FOAL

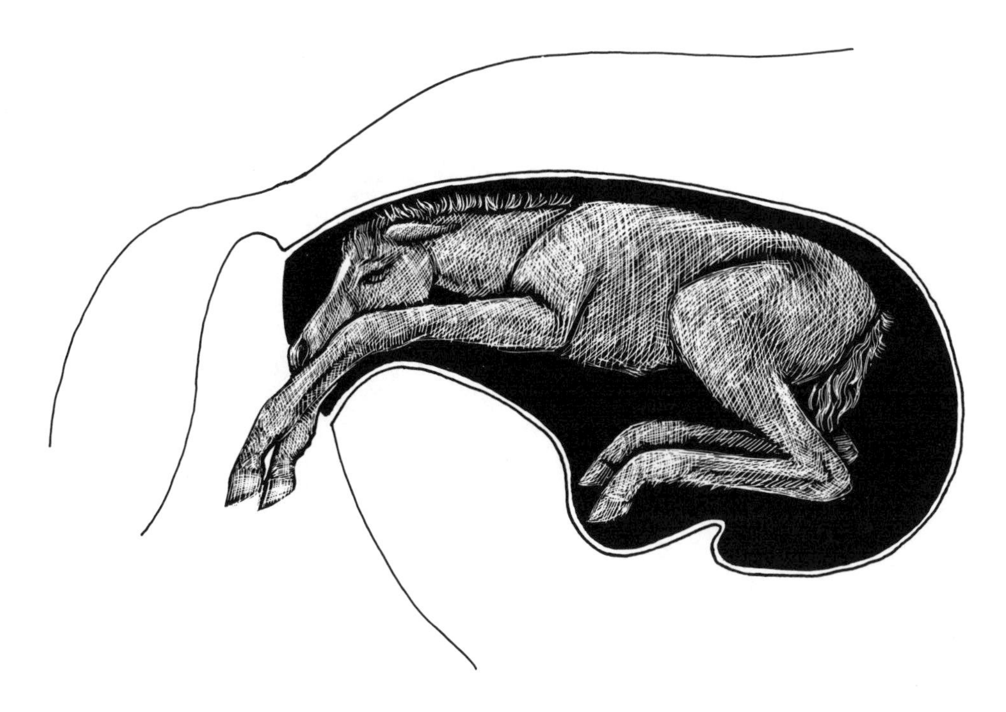

Normal anterior

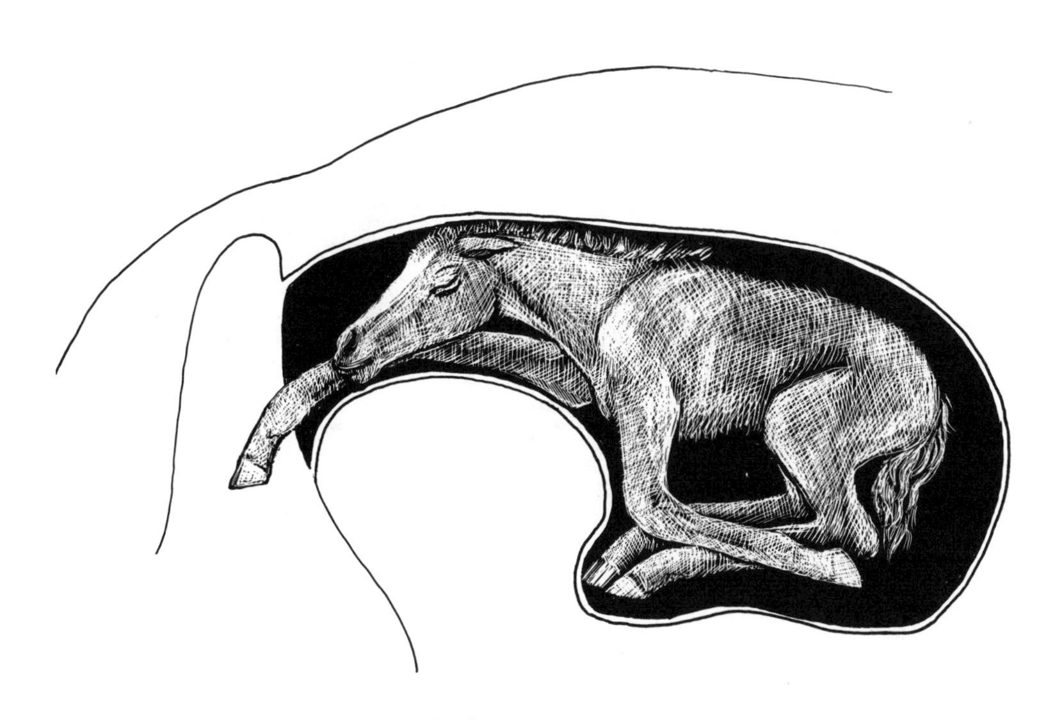

Left foreleg retained

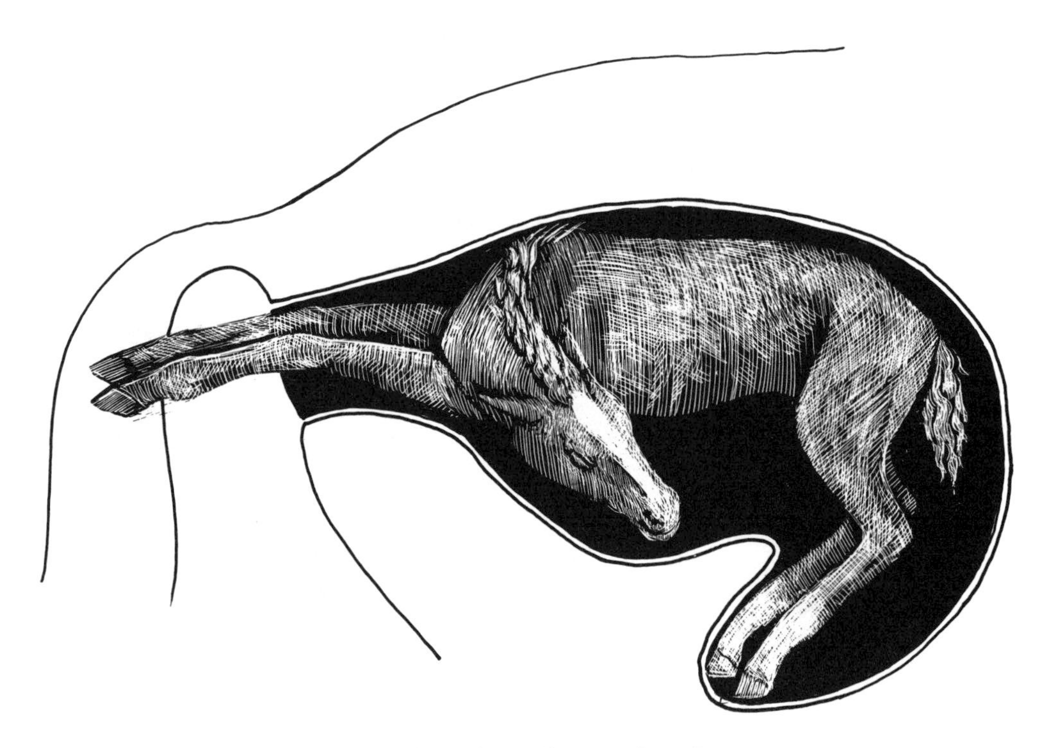

Head and neck retained

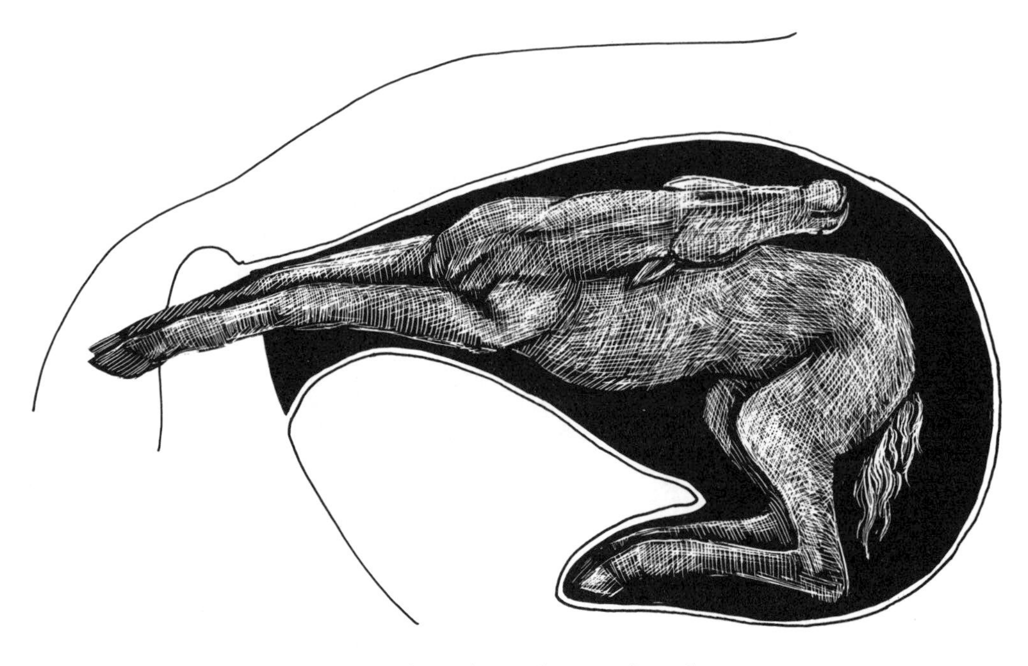

Head and neck retained

first foot and leg back a bit while you pull the inner one out toward you. Just getting the mare on her feet may accomplish this. If you find the foal's nose, but not the missing leg, you have a serious problem and probably had better call your veterinarian for help.

If you can't get help, wash up again and, using plenty of soap and/or lubrication, push on the nose with one hand and between the mare's strains reach in and try to find the missing leg. Be careful—the mare's uterus is fragile and easily torn. If you are careful and can grab the tiny hoof in your hand, bring it inward and upward under the foal's neck and over the pelvis. Once you have both legs, and find the head in the correct position, you might as well start pulling.

Pulling. To pull, put chains or nylon straps on *above the fetlocks* and pull down, putting on most pressure as the mare strains. Usually she will lie down about the time you start to pull. As you pull check to make sure the foal's nose is in the right place and that in bringing the leg up you have not pushed the head back too far.

The above directions for pulling may also be followed when the foal is coming in a normal position, but is just so big the mare can't seem to push it out unassisted.

NOSE PRESENTING

Both legs back and the nose showing usually means an already dead foal and unless you are experienced you had better call for veterinary help immediately. This position is often the position of a dead first twin. The second twin could be alive.

BOTH FEET PRESENTED AND HEAD BACK

This is the second most difficult common presentation after true breech. Again, unless you are experienced you had better call for help. If you can't get help and

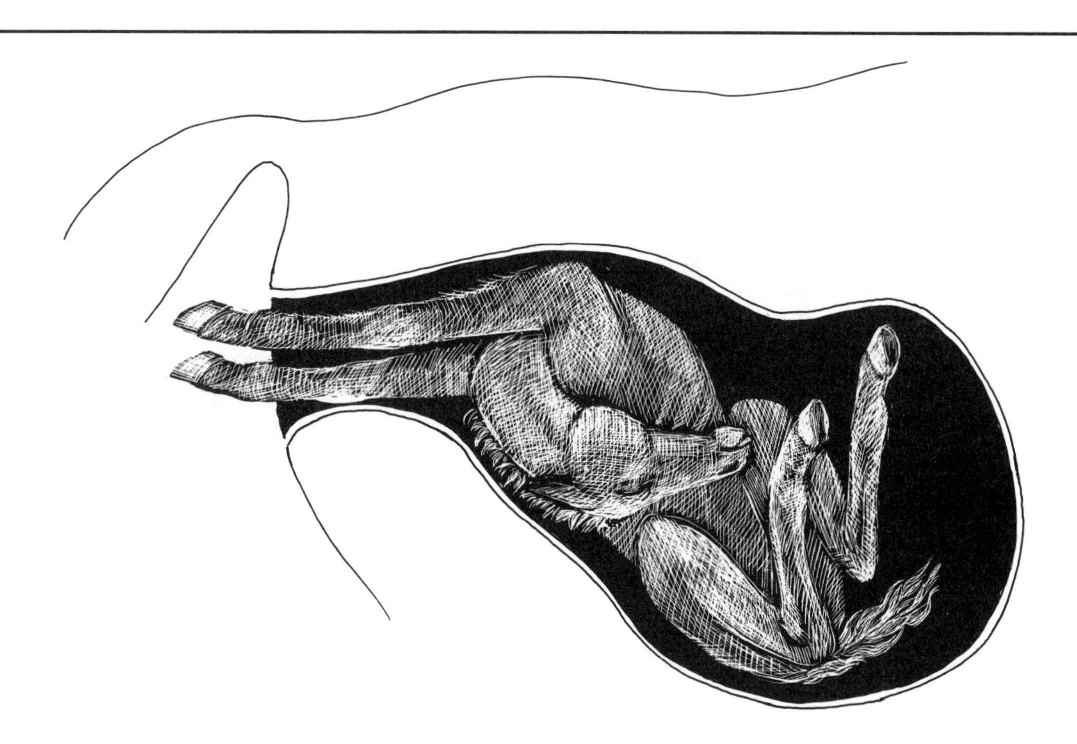

Head and neck retained. This can be confused with normal posterior presentation because of position of hooves.

are game to try, scrub and lubricate both your arms to the shoulders. With the mare on her feet, hold the foal back with one hand and try to find the head with the other hand.

If you are familiar with calving you may wonder why I say that most difficult malpresentations in the foaling mare end up with a dead foal. The problem is that the placenta releases from the uterine wall so quickly. The mare has a smooth placenta, unlike the cotyledons (buttons) in the cow, so that it breaks loose during foaling. There are exceptions, and you have nothing to lose by carefully trying to do what you can, but you must always be immaculately clean when working inside a mare. Although you may lose the foal, if you do no damage to the mare you always have a chance of a normal healthy foal next year.

It may be down directly between the foal's legs, but more likely is to one side. If you can get a finger or fingers in the foal's eye sockets or lower jaw, or a hand over his nose, you may be able, with a little luck, to pull the head up and into position. If you end up with a live foal in this situation it is a miracle.

Even if you see two feet coming always check to see if they are soles up or down. The foal's foot has a soft plastic-like yellow pad covering the sole, which shrinks away soon after birth. If this is up it either means the foal is coming out front legs first, but upside-down, or it may be the two hind feet, which is normal posterior presentation. If it is the former you may have a serious problem, as it might mean the head is back, too. Or it may be corrected simply by getting the mare up and down a few times.

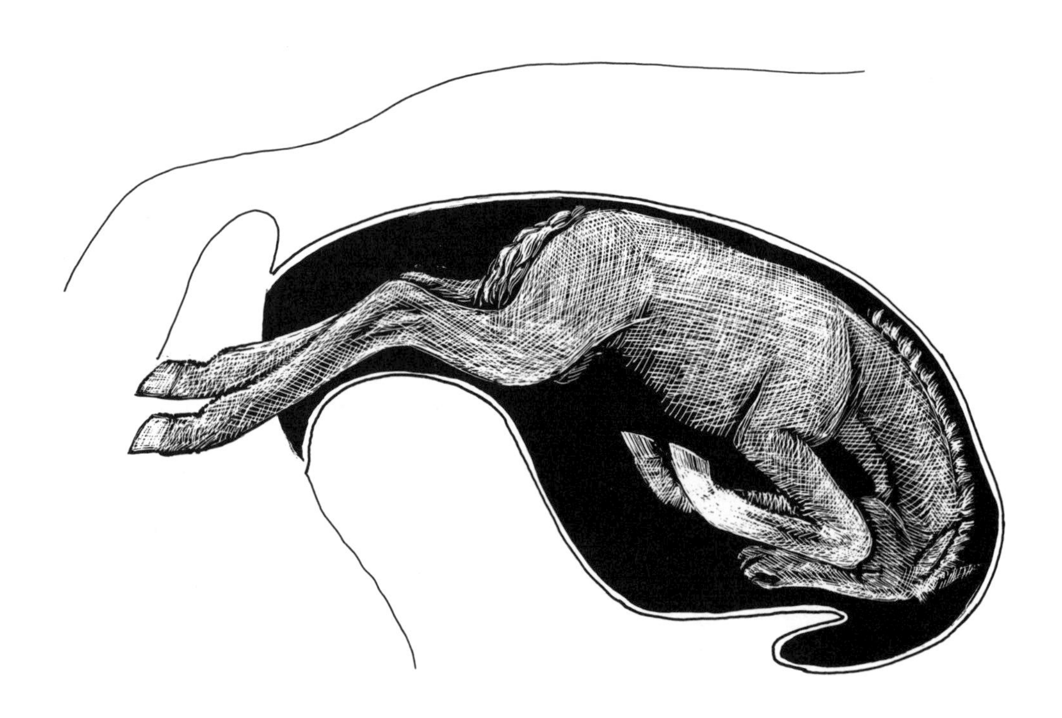

Normal posterior—not to be confused with breech

TURNING THE FOAL

A small foal can be turned fairly easily in a mare, but a large one can be a real problem. If at all possible get help on this one. If you can't you have nothing to lose by first examining the mare to determine whether you have front legs or hind. If you find the head in a near-normal position except upside-down, or off to one side, simply grasp each leg at the cannon and try to twist them so that the whole foal turns to its normal position. You still may have to reach in again and guide the head up into its normal position above and between the foal's forelegs.

POSTERIOR PRESENTATION

If after finding the soles up you reach in and find hocks it is a normal posterior presentation, not a breech. Normal presentation or not, you have a problem. Keep the mare on her feet as long as possible and try to get someone to help you pull. Although a mare can deliver a posterior presentation foal by herself she normally stops to rest for an instant as the hips go through. When she does, the navel cord is cut off by the foal's chest. When it tries to breathe it takes fluid into its lungs and suffocates before it is born.

If you have help, hook the obstetrical chains or nylon straps around the legs, *above the fetlock*. Once the mare starts to strain in earnest, pull down with all your strength. As the hips come through keep pulling even if the mare stops pushing. Usually in a posterior presentation the cord snaps as the foal comes out so there is no advantage to leaving the foal to lie quietly by the mother as in normal anterior presentation.

Assuming you have help, hang the foal upside-down using the obstetrical chains or straps. Clean out its mouth and nostrils and tickle the nostrils with a straw to induce sneezing. Slap the foal's ribs and if it doesn't

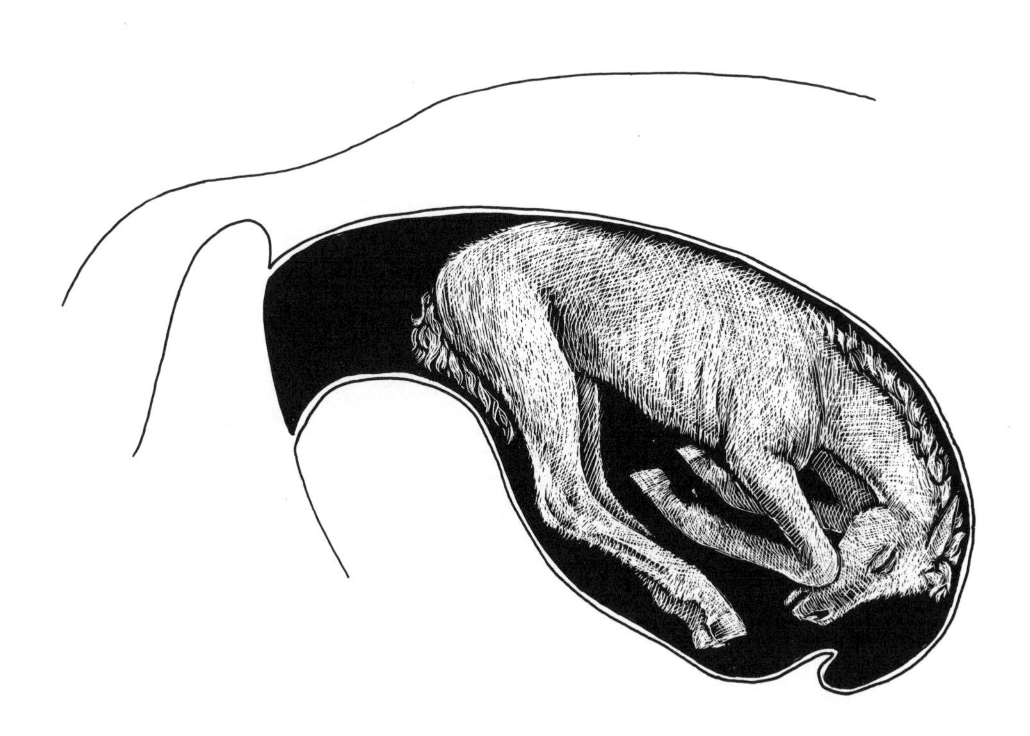

True breech presentation

start to breathe when everything seems to have run out, lower it and pump its chest rhythmically. If you have oxygen available try to get it started. Blow it into the foal's nostrils with the mask over the nose until it is breathing normally. Without oxygen, you may try blowing into one nostril as you close off the other and the mouth. This often works on a foal born head first, but on one that has sucked fluid into its lungs it may only make things worse unless it has been drained of fluid first. Still, if the foal isn't breathing you have nothing to lose by trying.

BREECH

If you find that you have a mare with a true breech, the tail coming first and both rear legs retained, you should call your veterinarian for professional help. If you can't get help and are game to try, after the usual scrubbing up, with the mare standing, reach in and push the foal's buttocks forward with one hand while you reach for a hock with the other. This is one of the most difficult maneuvers in a foaling case, and in a big Thoroughbred or draft mare it seems next to impossible. Sometimes the only solution is a Cesarean.

But if you can flex (bend) a hock so that you can grab one leg at the cannon, bend it inward and pull it toward you at the same time, you may be able then to slide your hand down to grasp the hoof and, shielding the uterus with your hand, draw the leg back to a normal position. If you can get one, you generally can get the other. A live foal under such circumstances is a rarity, but it can happen.

WORST CASES...

An unusual, somewhat similar position with the foal coming head and front feet first, but with one rear leg pointing forward and stuck in the pelvis, is probably the most difficult situation to get into, next to a rotated transverse pregnancy. It is usually made

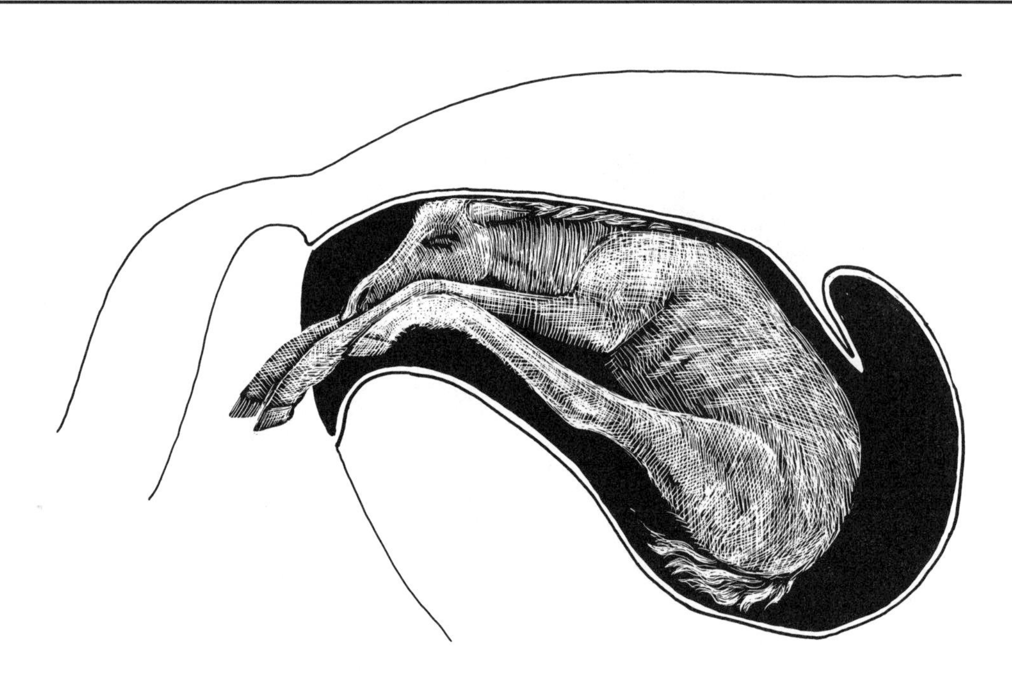

Hind limb deviation—"dog sitting position." If this presentation is not recognized and corrected immediately it will develop into a nearly impossible delivery.

worse by what appears at first to be a normal presentation being pulled until the leg is jammed and as stiff as if no joint existed. Your veterinarian may be able to correct it by using some old cowboy trick like pushing the leg back with the handle of a long-handled twitch, but often an embryotomy (cutting the foal out in pieces) or a Cesarean is the only answer. This is a rare malpresentation but I feel you should be warned that such a situation can occur. The first indication of it is when a normal-sized foal in a big mare cannot easily be pulled. To repeat, unless you are extremely experienced you must have veterinary help on this one, the sooner the better.

The other position I mention as being the absolutely worst is the rotated transverse pregnancy. This is usually seen only in big draft mares, if at all. It is a case where the foal is crossways, with part of its body in each horn of the uterus, and the uterus rolled so the foal can't be reached. A Cesarean, of course, is the only solution.

...AND MIRACLES

For most people, particularly with their first foaling, a veterinarian should be called at the first indication of trouble. Most of what I have described is not for the first-timer to attempt. Over the years, however, I have seen people inexperienced with foaling do some things that even they themselves could hardly believe. In each case it was because they were faced with a situation that required help right then, and although they called for help it needed to be instantaneous.

The case that amazed and pleased me the most was what I found when a young brood mare manager phoned to say that a mare was foaling with the feet showing from the vulva, but the head was coming out the mare's anus. This is all too common in the Thoroughbred and although the foal is usually pushed out alive by the mare, the anus and vulvar opening are torn into one big hole. If the mare survives she needs hours of surgery to correct it.

In this case, this young woman and the night man pushed the head back in and helped the mare deliver through the normal opening. When I arrived they were proud as could be of the lively foal almost ready to stand. And miracle of miracles, the hole between the roof of the vagina and the floor of the rectum healed by itself in a month. When faced with a bad one, call your veterinarian, but if your intuition tells you that you can do something you have little to lose by trying.

Post-Birth Care

Now if all the above has not discouraged you from ever breeding your mare, let me again reassure you that forty-nine out of fifty foals are born alive and well unassisted. So what do you do after your mare has a normal foal and everything has gone normally?

First, if the foal is lying with its rear feet still in the mare and the cord has not yet broken, let it lie and don't make the mare get up. The placenta is full of blood that, as the uterus contracts, is pumped into the foal's system.

Before we go any further, a word of warning. Certain individual mares, often the most quiet and affectionate, become extremely protective of their foals the first few hours after birth. Always keep one eye on a mare the first time you go into the stall after foaling to be sure she isn't going to attack you as you bend over the foal. Young children and dogs should never be allowed near a recently foaled mare.

Once the cord is broken, which should be done by the movement of the foal, pour tincture of iodine over the stump. If the foal is already on its feet put the iodine in a paper cup and dip the navel. In the rare instance when the cord does not break naturally, you may first try pulling *gently* with one hand while you hold the cord close to the foal's body with the other. If this does not work, break it using two hemostats, one next to the body and one two inches out.

As a last resort, cut the cord with scissors two inches or more from the foal's body. Usually there is a narrow section of the cord you can identify to see where to cut. There will be some bleeding, but it should stop in a few seconds. If not, tie the cord near the foal's body using a piece of muslin bandage or clean string. Then apply, or reapply, tincture of iodine to the stump, soaking the string as well as the navel.

If the cord has broken naturally and continues to bleed you may have to tie off, or in rare instances, hold the stump by hand until hemostats or clamp can be applied.

If it is cold in the stall take a big Turkish towel and rub the foal dry. If you need a heat lamp for the foal hang it high enough so the mare can't bump it, and be ever aware of fire danger.

While the foal and the mare are lying quietly is a good time to offer the mare a pail of lukewarm water (about 85°). Sixteen to twenty quarts of warmed water are sometimes taken by a mare at this time. Contrary to rules for "cooling out" when we restrict water, this seems to help the mare to replace fluid lost at foaling and allow her to clean more rapidly. By no means should the mare be allowed to gorge herself on cold water after foaling.

Foal Enema. On most large breeding farms all foals are given an enema immediately after birth. For the single mare owner working alone this is not easily done and is not always necessary. It probably is needed within the first twelve hours of life if it is early spring, your mare has not yet had grass, and the foal is a male. While the foal is still down one person alone can give a Fleet enema easily, using a human adult size. Better yet, you may use an enema outfit dispensed by your veterinarian and made especially for newborn foals. Warm soapy water and a soft rubber syringe or human enema outfit may also be used, but are difficult to use alone.

NURSING

If the foal is standing when you find it, you should stay as far away as possible, other than putting iodine on the navel, and not let any strangers, dogs, or other horses bother the mare until you are sure the foal has nursed. The biggest reason foals don't nurse is because we try too hard to help them. Most will find a teat and start on their own if left alone. When a mare kicks and won't let the foal nurse, usually the case with a mare's first foal, try tying the mare first. The use of a nose clamp or so-called humane twitch is often a great help with this type of mare, and tranquilizing her can also be effective.

Colostrum. If the foal has not gotten colostrum, the first milk, after five or six hours you must milk some out by hand and feed it with a dose syringe. Try to get in at least a pint over a period of an hour. If the mare doesn't have milk, contact your veterinarian and try to find a neighbor who has some frozen colostrum stored.

The reason why colostrum is so important is that this first milk contains antibodies the foal needs to prevent disease, and its stomach can only absorb the antibodies during the first few hours after birth. If there has been a delay of more than six hours in getting colostrum into a foal your veterinarian can test its blood for antibody content and, if necessary, give serum containing antibodies. Though expensive, this is necessary if you wish to raise a healthy foal.

If you did not give the foal an enema during its first hour of life, once it has nursed you should consider doing so, though you probably will need help. The foal's rectum is very fragile, and fatal injury can occur if you try to hold the little animal down with one hand or sit on it, and try to give the enema one-handed.

PLACENTA

In the meantime, the mare should "clean," or drop her placenta (cleanings or afterbirth) in less than three hours. If not you should notify your veterinarian. Retained placenta in the mare is serious and needs veterinary attention, not necessarily to have it removed by hand, as used to be done, but to give medication to prevent infection and the usual sequel of laminitis.

As soon as you see the placenta has been dropped normally, remove it from the stall and spread it out on a clean floor to see if it is all there, and a small patch is not left in the mare. Spread out, the placenta should resemble a child's Dr. Denton sleeping suit with one leg shorter and smaller than the other, about the size of a 100-pound grain sack. If you have any doubt that it is all there, save it and call your veterinarian for advice. For that matter, most veterinarians appreciate a call after a mare has foaled and all is well, so they can cross it off their "due" list.

On many breeding farms every placenta is weighed. Between ten and fifteen pounds is normal, with over or under weight said to be indication of disease. Your own veterinarian can best advise you if weighing is necessary for your particular situation.

If you were present when the mare's "water broke," or you watched closely as the foal came, you may have seen yellowish brown objects that appeared to be feces but were not. These are called hippomanes and are normal. I mention them so you will not be concerned the first time you see them.

INJECTIONS

If your mare had an injection of tetanus toxoid, and a booster a month before foaling, no injections are needed for the foal or mare. Until recently, if mares had not had toxoid it was suggested that both receive antitoxin to prevent tetanus. Now some veterinarians use toxoid in these cases instead. Depend on your own veterinarian for advice for your situation.

Just a few years ago a newborn foal, and sometimes the mare, was given a whole battery of injections, from antibiotics to vitamins and iron. Most authorities now feel the stress of the injections for the foal is worse than the diseases they are trying to prevent. However, if you have to reach into a mare at foaling, or if the foal is weak or appears anemic (with pale mucous membranes) your veterinarian may decide some medication is in order.

As soon after foaling and cleaning as possible, check your mare to see if she ripped badly at foaling and may need sutures. Mares that had a Caslick done during the previous pregnancy are sometimes sutured again immediately after foaling, particularly if it is the intention to breed them back the same year.

Watch your new foal, particularly if a colt, for indications it is urinating. Male foals, and occasionally females, may have a bladder ruptured at birth. Little indication is seen until the foal is two days old. If this is diagnosed, your veterinarian can perform surgery to correct the condition and save a large percentage of these. The alternative is death.

If you are like most horse owners, after your first mare's foaling you will have a great sense of elation. After forty years of foalings I still have the same feeling, and hope you always will.

CHAPTER XVII

AMERICAN QUARTER HORSE ASSOCIATION

A FOAL'S FIRST TWO YEARS

In ***the middle of a village*** *near my home, where a dairy barn once stood, is a small fenced pasture generally occupied by one or two Standardbred horses. If you drive down the main street of this village in early morning, just as the first light breaks over Alander Mountain, you will come upon one of these horses being driven in a "jog cart" down the street to Empire Road, past the Grange Hall to Empire Farm, to be worked on the half-mile training track there.*

Recently, instead of the usual two dark bay geldings, two mares, one chestnut and one grey occupied the paddock, each nursing a foal. Inquiring about the origin of the grey, which was obviously an Arabian and not a Standardbred, I was told, "That's Shadow, don't you remember her? The one whose mother abandoned her. Leigh raised her on milk-replacer."

"Did she let her foal nurse like that right off?" I asked.

"Half hour after the foal hit the ground he was up nursing and never stopped."

Noticing the size of the foal and guessing he was by a Standardbred stallion from his size and conformation, I agreed. "Looks like his mother is a better mother than his grandmother was."

"And then some!" was the answer.

The story of Shadow's birth and raising came back to me as one of the most serious foal rejections I'd ever seen. The size and beauty of this grey mare reminded me that Leigh, her owner, had certainly done a good job of raising a rejected foal on milk-replacer.

One fall Leigh had purchased an Arabian mare that turned out to be pregnant by an Arabian stallion, with the foal not due until September. The mare was wormed and given her shots, and appeared to be carrying normally.

The mare had twenty acres of old heifer pasture to run in with another mare and a gelding. Once a day they came up for grain, but the rest of the time, as winter led into spring and a long hot summer, they were off living the life of wild free horses.

As the supposed due date came an attempt was made to confine the pregnant mare to a smaller lot. She made such a fuss, however, that she was let out again with her "buddies."

One day a call came in that the mare had apparently lost the foal, since she had bagged and then appeared one morning with her belly gone, milk dripping from her teats. When she was caught and examined it was evident that she certainly had foaled, but there was no indication that she had been nursed.

Searching the pasture we found a placenta but no sign of a foal. If the foal was dead it should be near here, we reasoned, but a search through the alders and bog land nearby revealed no clue.

As always in such a case someone came up with the theory that coyotes or wild dogs must have dragged the carcass off, but although I've heard such stories a dozen times, and know that wild dogs in the area kill calves, they always leave some trace. Finally we gave up and continued on calls, leaving others to carry on the search.

After a few hours everyone else gave up the search, too, except the retired farmer who had pastured the land until it was sold as a building site. He had looked for lost calves in this pasture and said he just knew that, although we found no tracks or other evidence, the foal was there and alive.

Instead of walking through the bushes as we all had done, he went to a high spot in the pasture and sat down. His theory was that this foal, although only hours old, was wild as a deer and kept running ahead and out of sight as we looked. Sure enough, an hour after he ascended his watching post he saw a chestnut foal appear, not only acting like a deer, but looking like one, too. (Grey Arabians are often chestnut color at birth.)

Several people surrounded the alder patch the foal was in. It was caught and brought up to the stable and confined in a small high-fenced yard with its mother. When it approached her she rewarded it with a wide swinging cow kick that could have killed a bear, but foals at that age are tougher than bears. The mare was caught and twitched, but still she squealed and kicked at the foal. Tranquilizer was tried with little effect. Finally, with just plain bull strength and determination, four of us held the mare while Debbie, my veterinary technician, guided the foal to a teat. It nursed with relish.

Two hours later an attempt was made again to let the foal nurse, but the mare kicked even more violently. The foal was removed while we tried to think where we could get colostrum at that time of year. Out of curiosity, and perhaps foolishness, I attempted to feel the mare's udder, then to take streams of milk out by hand. She stood like an old Brown Swiss cow enjoying being milked. I milked a pint from her and the foal was bottle-fed. Another attempt was made later to bring the foal in with the mare, but again she would not let it nurse. Yet if the foal was not around, the mare would let anyone milk her by hand.

For three days the mare was milked every four hours and the foal fed. By that time Leigh had had enough, and started the foal on milk-replacer.

For a few months I saw the filly, which was named Shadow, quite often. She seemed to be growing as well as any foal, in fact, far better than one would expect considering her September birth date. As a yearling Shadow was moved to another location with the rest of Leigh's horses. Until I saw her nursing her own foal I had forgotten about her.

I have seen many starts at raising orphans or rejected foals over the years, with poor to fair results on most of them, but an occasional successful attempt. The results with Shadow proved to me that they can be raised successfully. All the tools are there to be used, but few people have the determination to make them work.

R**aising a foal** to a well-grown, well-trained two-year-old begins when it nurses for the first time. Although it does not always require the determination of a Leigh, it does require some. As in so many other things, if it is worth doing, it is worth doing right.

That first feeding of colostrum is so important that if it cannot be accomplished naturally, artificial means must be used. Although I understand that in modern cattle practice calves are force-fed colostrum with a short stomach tube in order to be sure it gets inside them, in the foal a stomach tube should be a last resort and used only by a veterinarian or experienced technician. If a foal will not or cannot nurse, small amounts of colostrum, two to eight ounces fed every two hours with a bottle and nipple or a catheter-tip dose syringe, or even sucked by the foal out of a pail or pan, are worth far more than one big dose. The foal's stomach is a tiny thing compared to a calf's of the same weight, and it needs smaller amounts given more often. Not only is the colostrum important to impart immunity in the foal, but it is also a rich source of fat (14 percent), protein (20 percent), and sugar (50 percent) to give the foal energy and keep it warm.

USING MILK-REPLACER

If you must raise a foal on milk substitute, the first three days it should still get colostrum. After that you can give Foal Lac made by Borden, available in most feed supply stores in horse areas, following the directions on the container. In an emergency, milk-replacer made for raising veal calves is satisfactory for foals. Its fat and protein content are similar to Foal Lac.

A foal should receive 10 percent of its weight daily in mare's milk or milk substitute. Thus an eighty-pound foal would need eight pounds (eight pints or 128 ounces) daily. The first day you may need to feed a weak foal every hour, thus four to six ounces every hour is enough. This can be stretched to every two hours using eight to twelve ounces and at three days can go to every three hours on a healthy foal, using a pint or sixteen ounces every feeding, totaling eight pints a day.

USING A NURSE MARE

Another way to raise orphan or rejected foals is on a nurse mare. If you have an orphan foal during peak foaling season there is a good chance that someone within 50 miles has a mare with a dead foal. In the racehorse breeds this is often done instead of using milk-replacer since the foal's personality and character develops better if it starts life following a mare around instead of a human.

Getting the mare to accept another foal is sometimes difficult, but can be done with patience. Some people keep mares just to rent to Thoroughbred owners to raise expensive well-bred foals on. This permits the Thoroughbred mare to be sent away to be bred without the foal accompanying her. Of course, this means either destroying the original foal or raising it on milk-replacer. There is not space here to go into all the fact and lore of nurse-mare use, but one tip I learned by accident is that if you can *bring the foal to the mare*, acceptance is more likely than the way it is more often done, taking the mare to the foal.

Raising a Normal Foal

Let's assume, however, that your foal is nursing the mare just fine and everything is going the way it was intended to. What is the standard procedure in raising a normal foal?

Beginning at twenty-four hours of age or younger, put a small foal halter on the foal and take the mare and her offspring outside for fresh air and, hopefully, some sunshine, even if only for an hour or so. Only sheer ice or cold rain should be allowed to prevent this. You will notice that in winter if the mare has been out in the cold her foal will be born with long hair. In warm weather foals are born with coats short as velvet. Mother Nature equips the foal to go out in the cold, and it may do so a little longer each day.

When you bring the mare and foal back in remove the halters from both. The first few days the foal will follow the mare and will continue to do so for months with no control necessary. If you have put the halter on and taken it off once a day, however, the foal will be used to it by the time it is a week old. A light lead can then be put on the foal. Lead it outside with its mother in the morning and in the afternoon catch it and lead it inside.

HANDLING A FOAL

To catch a foal not yet halter-broken, approach it from, and facing, its left side. With your left arm under its neck and ahead of its shoulders, and the foal's tail held up straight in your right hand, almost lift it off the floor and force it gently to stand still. Don't do this in a rough way — let the foal think it is a big hug, but let it know you are the stronger one.

A breeder of Belgian draft horses once told me, "I pick my foals up when they are twenty-four hours old and won't put them down until they hold still. The rest of their lives they think I can still do it." I believe he is right since it was always a pleasure to work around his horses. Despite breed differences in temperament the foregoing method is as effective on Thoroughbreds and light breeds as on draft horses.

FEEDING A FOAL

At a week of age foals will start nibbling on hay with their mothers. Be sure it is of good quality, such as early first-cutting alfalfa-grass hay of good color, free of mold and dust.

At about nine days, plus or minus three, most foals will have a bout of diarrhea, but won't act sick. Since this seems to coincide with the mare's nine day "foal" heat most people believe it is from changes in the milk at that time. It has been said that this is also the time the foal has the first batch of strongyle eggs hatching in its digestive tract. I don't know which is the correct answer, but I do know that if the diarrhea lasts more than three days, giving large doses of thiabendazole wormer seems to stop the diarrhea.

Soon after its first week your foal will start to nibble grain that falls from its mother's manger. The sooner the foal eats hay and grain the better. By three weeks of age you may start to give the foal its own grain in a manger it can reach. To make it easy, use one of the complete ration pellets. Foals will not overeat on these, but if the grain accumulates in the manger and becomes soiled from flies and slobber from the foal, throw it out, clean the manger, and start over. If the mare steals the foal's grain, and she probably will, put it behind a bar too low for her to get under and too high to reach over.

Depending on the season, foals can be on pasture with their dams from birth on. Even a foal born in March in the northern United States or southern Canada can go to pasture with its mother by mid-May. Mares nursing foals at pasture, no matter how good the grass appears, should receive grain and some dry hay, too. Salt blocks should also be available. At pasture you should build an enclosure (creep) that will admit the foal but not its mother, where the foal can get grain and sometimes hay, too. Halter and tie the foal once a day, when you give it grain. Or, if you have the time, halter and tie both mare and foal while they eat their grain, as an alternative to building and using a creep.

A moderate (14 percent) to high (18 percent) protein grain or one of the "complete" rations is satisfactory for foal use. There are also commercial grain rations especially for foals, which can be fed according to manufacturer's recommendations. For example, at five weeks a foal needs one pound of grain per day, at eight weeks two and a half pounds, and by twelve weeks this is increased to seven pounds. Unless the foal is a "hay hog" it should be fed good-quality hay free choice.

HEALTH CARE

Mares and foals should be wormed one to two months after foaling. Your veterinarian can recommend when, and what wormer is best suited to your area and situation. At the same time the foal should receive its first tetanus toxoid.

At one to two months of age foals should have their feet trimmed by a skilled farrier. This early trimming can affect their conformation and the soundness of feet and legs the rest of their lives. Almost as important is the early training the foal receives in becoming accustomed to pick up its feet to be trimmed.

The farrier will try to trim the foal's feet to maintain or achieve even contact of three points, inner and outer heel and the toe of each foot. The hoof angle, as in a more mature animal, should correspond with the fetlock angle.

At three months of age the foal should have a tetanus toxoid booster and be started on other immunizations. The usual ones are equine influenza, rhinopneumonitis, Eastern and Western equine encephalomylitis. Your veterinarian may suggest others needed for your particular area. Worming should be done on the advice of your veterinarian.

WEANING

The longer a mare milks after the first ten days the more water and the less nutrients her milk contains. There is apparently a great deal of difference from individual to individual. Some mares that appear to have the most milk raise the poorest-growing foals for this reason. The foal drinking large quantities of this watery milk has no room for hay and grain. If a worm-free foal is not growing as it ought to, one should consider weaning it and getting it on a more nutritious diet of hay and grain. Foals can be weaned as young as one to two months with good success. Milk-replacer pellets are available to supplement the grain feeding.

Whether weaning is done early, at one to three months, or later, by six months of age, it should be done "cold turkey," for the good of both the mare and the foal. For best results the foal should be separated from the mare, far enough away so that one cannot see *or even hear* the other. Both, particularly the foal, should be in a tight-walled or fenced stall or paddock which it cannot climb on, over, or through and become injured. If at all possible, two or more foals should be together. If another foal is not available, a gentle gelding, a yearling filly, or older mare that has not recently been nursed can be used to keep the foal company. The company of a pony, goat, or even a dog is better for a foal than having it completely alone.

Mares at weaning time should be confined to stalls or grass-free paddocks, fed dry coarse hay with little or no grain, and have their water consumption limited by hand watering. Their udders will fill rapidly immediately after removing foals, and although they should be watched closely for signs of trouble, they should not be hand-milked to relieve pressure since this encourages milk production. Symptoms of trouble in the mare related to weaning are colic, depression, fever, and, on rare occasions, "hives," which manifests itself with itching and swellings on various parts of the body, particularly body openings. If these symptoms appear, milk the mare by hand and, of course, call your veterinarian.

After two weeks, if the mare's udder has softened and is shrinking she may be turned at pasture with other non-nursing mares, but should receive less grain than while nursing.

Once recently weaned foals are accustomed to being away from their mothers they should preferably run at pasture with other foals. Any other non-nursing horse, except a stallion or uncastrated yearling, is better company than none.

Never turn a foal back with its mother until they have been apart at least a year. Some of the most "spoiled rotten" and dangerous horses I've worked on are mature geldings and mares that still follow their mothers around.

Weanling foals are often the most neglected animals on the farm. They must be watched continually for proper foot growth and their feet must be trimmed every six weeks, or more often if improper conformation is noticed. Worming should be routine and follow-up immunizations should be done on schedule.

FEEDING WEANLINGS AND YEARLINGS

The toughest problem with weanlings and also yearlings is to keep them growing, without growing so fast that they develop what some describe as "leg bones growing faster than the tendons," or contracted tendons. This seems to be a particular problem with Thoroughbreds and Quarter Horse breeds, although it can be seen in any breed. An imbalance of calcium and phosphorous is blamed for this in some cases. In areas of the country where too much lime is applied to hay fields over the years, hay, particularly legumes, contains a high amount of calcium. If you notice that your weanling or yearling is wearing its toes down too fast and/or appears to have a stiff gait, try cutting back on its grain. Check to make sure you are not feeding a mineral mix high in calcium meant for dairy cattle, or that such a mineral mix hasn't been added to the grain at the factory. You might even have to change from high quality alfalfa hay to broome or timothy. The addition of dicalcium phosphate to the ration sometimes gives good results.

The above is a direct contradiction of publicity prevalent in past years that most rations are deficient in calcium. Rather than try to cover all the possible problems of mineral metabolism in weanlings and yearlings I shall only caution you here that what is correct in one area may be incorrect in another, due to soil content of the area where the hay and grain were produced. This is true not only of calcium and phosphorous, but of trace minerals such as selenium and cobalt as well. Your best source of advice is your local veterinarian and/or equine extension agent.

Yearlings need less protein than foal's on creep feeding or weanlings. All three age groups (creep, weanling, and yearling) need more protein if fed grass hay than if fed legume, since good alfalfa contains so much protein itself.

The best advice I can give you is to find out what others who raise horses successfully in your area are feeding and follow their lead. Even then, don't be afraid to change if an animal is not growing well, is thin, is growing "too fast," or is too fat.

GROWING UP

Uncastrated colts should be separated from fillies by the spring of their first year. In the non-racehorse breeds castration as early as the spring of their first year, which means eight to twelve months, is not too soon.

Weanlings and yearlings should preferably run with others of their same age summer and winter, with an open shed for protection from cold rain and hot sun, and plenty of space to gallop and play. The person with one weanling or yearling may have to substitute horses of other classes (geldings, open non-nursing mares, etc.). Often neighbors who have a single yearling or weanling are glad to share space so two or more can run together. I have known owners with a single colt or filly who go out and buy a second of the same age and sex so their single animal can develop normally.

Again, on yearlings as on weanlings, don't forget to keep up the worming, and at a year or the first spring, immunizations used in your area should be repeated. Hoof trimming, as mentioned, must be attended to every six weeks.

Once a week, or more often if you have time, each yearling should have a "school lesson." This need only be a short time on a lead line, and perhaps being taught to lunge. Yearling horses, like young children, have very short attention spans, and lose interest in whatever they are being taught in as little as ten minutes. A reward of a carrot, apple, or other treat will keep them eager to please.

The first fall of a yearling's life it may be taught to wear a bridle and saddle or harness, but hold off on any actual riding or work until it is at least two years old. Give it a chance to grow up.

CHAPTER XVIII

CAROL DEMAYO/BONNIELEA FARM

WHEN THE END COMES

*"**Doctor,** this is Marsha Ashley on September Mountain," the sad voice said. "I just phoned Dr. Stevens's office and find he is no longer able to practice due to poor health. They suggested I call you."*

Mrs. Ashley went on to explain that Dr. Stevens had cared for her horses for thirty years and although over the years other veterinarians, including me, moved into the area and were closer, she had never considered calling anyone else.

Her immediate problem, she said, was Comanche, her oldest horse, who had developed a chronic diarrhea. She went on to explain that Dr. Stevens had warned her a year earlier that this thirty-year-old gelding was getting near the end of the road. His teeth were so badly worn that he could no longer eat hay, so he was fed a pelleted "complete ration" soaked in water. For a year he seemed to do quite well on this, but now was losing weight. She wondered if I could make a call to her stable, check out Comanche, and discuss her other horses. She added, "This is no emergency, so please come when you have time, plenty of time, to hear my problems."

Mrs. Ashley, her husband, and her now-grown children had been participants in gymkhanas and competitive trail rides for many years. Their hardware store in a neighboring town was a good business, permitting them both the financing and time to spend on their horses. I had seen them and their horses at local events and knew that their reputation for having the best horses in the best condition was well deserved. I sensed that there was more Mrs. Ashley would tell me about Comanche when I made the call.

When I arrived at the farm I was relieved to see that Comanche was not skin and bones. In fact, at first glance he appeared in as good shape as I'd remembered him years before when he was one of the top competitive trail-ride horses in New York and New England. Comanche was of unusual marking, being a true black, but having the white blanket and spots of an Appaloosa. Mrs. Ashley told me she had owned him twenty-eight of his thirty years, that he was half Thoroughbred, and that he had competed in trail rides until twenty-six years of age. Since then he had lived in retirement.

Further, she told me that Dr. Stevens had at first hoped the diarrhea was not due to diet or change in intestinal flora, but caused by tapeworms, sometimes the case in old horses. Everything including tapes had by now been ruled out and no medication seemed to help except a mixture of vitamins and bacterial culture given by mouth. As soon as this medication was discontinued the diarrhea started again and Mrs. Ashley had decided that, unless I had some other idea, she would like me to set up a time when I could put Comanche down.

Mrs. Ashley showed me copies of lab reports on tests Dr. Stevens had done and I agreed that there was nothing more that would be humane and practical to try on an old horse with no teeth. Obviously, as Dr. Stevens had suggested, Comanche had "reached the end of the road."

Mrs. Ashley told me that over the past thirty years she and her family had owned over twenty-five horses. Some had been kept for only a few months because they turned out to be unsuitable for gymkhana or trail rides. Some didn't get along with the other horses, or didn't compete well and were sold with no claims or promises for someone else to try. Some were bought "green," trained to be top performance horses, and sold to people who were willing to pay top price. Some, for whom the family had no strong sentimental attachment, were sold when they reached old age to a man who slaughtered horses for dog food.

"All horses, and all of us for that matter, have to die sometime," she said. "I do not feel it is cruel to have a horse put down for dog food, but it is cruel to sell them at an auction knowing full well that they are too old to be bought by anyone who would give them a good home. Those poor things end up going from pillar to post, unloved and uncared for until they die a slow, cruel death."

Plans were made to have a local con-

tractor with a backhoe dig a hole near a giant oak that Comanche used to like to stand under. My veterinary technician and I would come and euthanize Comanche, dropping him in the hole, while both Ashleys were away, and the contractor would fill the hole before they returned.

On the appointed day we arrived at the Ashley farm, finding Comanche waiting for us in his stall with his best halter on. On the box stall door was a hand-written note of instructions. He was given a tranquilizer and then led to the edge of the hole, where a second intravenous injection, actually an overdose of concentrated anesthetic, was administered. The bottom of the grave was already covered with one of the many blankets that Comanche had won and, as explained in the note, a second blanket to cover him with was left nearby.

As my assistant and I walked back across the field toward the barn neither of us spoke. We could hear the backhoe coming up the drive as we fastened to the stall door a note to replace the one Mrs. Ashley had left for us. It said, "Everything went without a hitch. We followed your instructions and Comanche is asleep under his favorite tree."

One of the most difficult tasks a veterinarian must perform is euthanizing an animal. Upsetting as it may be, however, it is one service for a client's animal that, if done properly, can give a feeling of an unpleasant job well done.

The biggest difficulty in euthanasia for horses is being able to bury the body. In rural areas, if you are willing to pay a contractor you can usually find one who can work his schedule to fit yours and the veterinarian's so as to bury a horse at once. The hole does not have to be dug in advance, depending on the owner's wishes. Sometimes the horse is put down and the hole dug next to the body. The horse is pushed into the hole by the machine.

The worst problem is in suburban areas where, even though zoning and health laws permit keeping live horses, burial of them is forbidden. In the North, of course, there can be the problem of frozen ground. If you have a horse that you want put down because of age start to plan by early fall to have it done. One of the hardest things on a horse owner is to try to get one more winter out of a faithful old horse, and to have it go down and be unable to get up during the cold winter months.

Few areas have rendering trucks that call regularly any more, and getting a dead horse hauled out of a barn and to an area where it may be disposed of, either by rendering or burial, can cost hundreds of dollars, plus the heartbreak. Whenever possible, plan ahead.

Sending a horse to the "butcher" may sound harsh and cruel. Yet horses that are used to being transported go on a van easily and in most cases are dead within hours of leaving their stable.

Few people who really love horses want to send an old horse to an auction where it might be bought by a person who figures on getting one year out of it at a summer "riding school" or camp.

Despite our American revulsion for eating horse meat, a good price for horse meat assures one that most old horses end up with a humane death, rather than being underfed and poorly cared for until they literally die of old age.

Before the situation arises when a horse has to be put down or moved because of age, discuss it with your veterinarian. Each veterinarian has his or her own method or technique for putting a horse down. He or she will explain these to you, but will let you make the choice. Each case is different, and the aim is to do what is least frightening to the horse. The advent of tranquilizers has made it easier to accomplish the job with little or no fright to the horse. In the extreme, a man-shy horse might better be put down by shooting, which may sound gruesome, but if properly done is the most humane method.

Despite the best planning, you should expect the unexpected. Should you have a horse die suddenly or be euthanized for some reason during a time of year when you can't have a hole dug to bury it, you may have to pay to have the body hauled to a rendering or dog food plant, or to a place such as a gravel bank where burial is possible. In such a situation your local veterinarian is the person most apt to know of a solution.

One must go on the supposition that quality of life is far more important than quantity. One must recognize the fact that some old horses' entire personality changes, and although they appear healthy they become mean and unmanageable. Plan ahead, and know that you have done for your horse the last kindness in a way you won't have to regret.

METTLER

EPILOGUE

*A **few years after** Sue and Jim Deane's sad experience with Nadia the following article appeared in the local paper.*

ATTORNEY SPEAKS AT STATE HORSE COUNCIL MEETING

Attorney James Deane spoke Saturday evening at the annual banquet of the State Horse Council. His subject was "You, Your Horse, and the Law." Attorney Deane, his wife Sue, daughter Jenny, and son Jamie, are horse owners and active in the local riding and 4-H clubs. He has had extensive experience in cases involving horses, horse owners, and the law.

Deane told the council members that just as prospective horse owners must learn all they can about horses before buying a horse, they must become familiar with laws covering stabling, health, riding, and driving horses in his local area. However, the attorney stressed, just as horse owners seek professional advice from farriers, trainers, veterinarians, and Extension personnel in caring for their horses, they must seek professional advice from an attorney familiar with the law regarding horses when they need legal advice.

Jim Deane has obviously changed radically in his attitude toward horse care. Owning a horse is a wonderful experience; it is also an enormous investment of money, time, worry, and love. If there's one lesson I'd like to leave you with, it's the very one Jim learned so painfully. Draw upon the expertise of professionals, whether they be veterinarians, horse trainers, breeders, farriers, or riding instructors. They have spent their lives learning things that will help you avoid many problems.

If you ask questions of the right people, and follow their advice, you will be the richer for it in every way.

GLOSSARY

The words and phrases in this glossary are either horsemen's terms for horses and horse equipment, or technical and medical terms defined in horsemen's language. The list is certainly not all-inclusive since there are many words and phrases peculiar to certain types of horses and to certain parts of the country which, if added, would have made the list nearly endless.

For further clarification of anatomical terms, see diagrams on pages 5 (horse) and 41 (hoof).

abscess—a localized collection of pus surrounded by inflamed tissue
acre—unit of measurement of land area, 4,840 square yards or 43,560 square feet
AI—artificial insemination, the mechanical introduction of semen into the genital tract of the female
alfalfa (*medicago sativa*)—a leguminous plant used primarily for hay, usually high in protein and calcium
anestrus—when a mare is not having or showing heat or estrus
artificial vagina—a mechanical device with a rubber liner used to collect semen from the stallion
ascaris (plural ***ascarids***)—large white intestinal parasite; in the horse the common ascaris is *Paracaris equorum*
bag—(n.) udder, mammary gland; (v.) enlarging the udder prior to foaling
bald—white or light color on a horse's head from poll to nose, including around the eyes
barium—a hard metal spot welded to the bottom of a horseshoe to help keep a horse from slipping
barley—a small grain similar to rye or wheat
barren—condition of a mare that did not become pregnant during the breeding season
bastard strangles—strangles that result in abscesses in internal glands
bay—a horse of brown or reddish brown color with black mane and tail
bell boots—bell-shaped rubber covering that fits over the horse's foot and hoof to protect from injury
bib—a device fastened under a horse's lower jaw to prevent it from chewing or licking itself, while still allowing it to eat and drink
bit—a device placed in the mouth of the horse as a means of control attached to the bridle and the reins or lines
black—a horse with a truly black coat with no brown or reddish hair
blanket—a fabric cover for a horse's body, usually made of wool or heavy material (see sheet); a marking of lighter color over the rump of a dark horse
blaze—white or light coloring on a horse's face, between the eyes from poll to nose
bloodworm—usually refers to strongyles
blow up—when a horse suddenly loses its temper
booster—a repeat immunization to restore or increase the amount of immunity
bot—the grublike larva of the bot fly found attached to the lining of the horse's stomach
bot fly—(*Gasterophilus nasalis, hemorrhoidalis* and *intestinalis*) a fly resembling a honey bee which, according to type, lays small whitish eggs on a horse's legs, nostrils, and mouth
box—boxstall, a four-sided stall to confine a horse
breast collar—a horse collar that fits over the horse's chest instead of around its neck
breechen—the part of the harness that fits over the horse's rump and holds the load back or permits the horse to back it up; also called "britchen"
breed association—the organization that registers the birth and pedigree of a particular breed of livestock
breeding shed—the building in which breeding takes place
brown—a horse with a truly brown coat, or more usually black with brown or reddish hair around mouth and nose
bulb of heel—the rounded portion of the horse's foot just behind the hoof
"bute"—phynalbutazone, a drug which reduces pain
calk—(n.) a pointed projection on a horseshoe to prevent slipping; (v.) to injure with the calk on a shoe
cannon—the area between the knee on the front leg, or hock on the rear leg, and the fetlock
cannon bone—the metacarpal on the front leg or metatarsal on the hind leg, the bone between the carpus or tarsus and the phalanges
cap—the remains of a "baby" tooth covering the permanent tooth
catheter-tip dose syringe—a large hypodermic syringe with a blunt nozzle tip
Caslick—the surgical technique in which the vulvar lips of the mare are cut and sutured so that they grow together making the vulvar opening smaller
cavesson—part of the bridle that goes over the nose and under the horse's jaw
cecum (*caecum*)—the blind gut; in the horse it is huge compared to other animals, holding five to ten gallons of ingesta
cervix—the narrow neck or mouth of the uterus
check—a strap that keeps the horse's head up
check rein—a strap that fastens to the bit to keep the horse's head up
"click"—breeding term for situation where certain blood lines, if crossed, produce exceptional offspring
clitoris—sensitive mound of erectile tissue in the lower portion of a mare's vulva
clover—a legume used for hay and pasture
coffin joint—the joint within the hoof of the horse between its short pastern bone (second phalanx) and the coffin bone (third phalanx) also including the navicular bone
Coggins—the laboratory test for equine infectious anemia, or swamp fever, developed by Dr. Leroy Coggins
colic—(n.) spasmodic pain in the horse, usually caused by spasm of the intestine; (v.) the reaction of a horse to abdominal pain, kicking, rolling, sweating
colostrum—the first milk, containing high protein, sugar, and, most important, globulins that impart temporary disease resistance to the newborn foal
colt—a male horse under four years of age; young racehorse of either sex the first year of training
common—an ordinary, plain-appearing horse
complete ration—a usually pelleted ration, containing all the necessary nutrients except water
cooler—a large heavy blanket, usually wool, to put on a horse after a workout
coronary band—the top of the hoof between hair-covered skin and hoof where growth takes place

coronet—coronary band
corpus hemorrhagicum—blood clot that fills pit on ovary immediately after ovulation
corpus luteum—yellow gland tissue that replaces corpus hemorrhagicum
cover—to breed a mare
cow-hocked—a horse with legs angled at the hock similar to a cow's
cradle—a device put on a horse's neck so it can't reach to bite or lick its sides or legs
cribber—(see **wind sucker**)
cross-tie—to tie a horse from two sides, with ties running from the halter to a wall or post on either side
culture—cultivation of living cells in prepared media—the technique used to determine if a mare's genital tract is infected
culture plate—a plastic dish containing jelled material on which bacteria will grow
curb bit—a bit with a U-shaped projection on the bar which pushes on the horse's tongue or roof of the mouth when reins are retracted
curry comb—a metal, plastic, or rubber device with many small teeth for cleaning hard-packed filth off a horse or cow
dam—female parent
dandy—a medium-hard brush for grooming to remove loose hair and dirt
dappled—rings or spots of different-colored hair on the coat of a horse
distemper—an old name for strangles in the horse, sometimes used to denote any infectious respiratory disease
DMSO—dimethyl sulfoxide, a solvent which, when applied to the skin, is rapidly absorbed
double tree—device that connects two single wiffletrees
draft horse—a horse used to pull heavy loads
dressage—horse show class in which the rider guides the horse through complex maneuvers with slight movements of hands, legs, and weight
Dutch collar—similar to breast collar
EEE—Eastern Equine Encephalomyelitis, viral disease of horses affecting the brain
E.I.A.—Equine Infectious Anemia, known as swamp fever
ejaculation—emission of semen from the stallion's urethra
elbow—joint between the humerus and the radius and ulna, located on the foreleg between the shoulder joint (*scapula humeral*) and the knee (carpal joint)
electrolyte—a water solution of salts used to replace or reinforce the normal salts of the blood
embryo—the early stage of development of the fetus
endoscope—an instrument using fiberoptics to view the inside of body cavities
ensilage (silage)—fodder such as corn or grass preserved by storing without air in a silo
equine—the family of Equidae, horses, asses, and zebras
equine encephalomyelitis—a viral disease causing inflammation of the brain and spinal cord
equine influenza—a viral disease affecting the respiratory tract of the horse
E.V.A.—equine viral arteritis
equine viral arteritis—a viral disease of the horse, usually mild but often causing abortion in the mare
estrus—"heat," reproductive period when mare will accept stallion
farrier—skilled horse shoer
feed bag—a sack usually of canvas and leather held on the horse's nose by a strap behind its ears allowing it to eat grain without a manger or other container
fermented feed—fodder preserved by storing in piles or airtight structures causing it to ferment and heat; also any feed that has become damp accidentally, causing it to ferment
fetlock—the tuft of hair on the back side of the fetlock joint
fetlock joint—between the cannon (metacarpus or metatarsus) and the pastern (first phalanx) including the sesamoid bones; sometimes referred to as ankle
fiberoptic—bundles of glass fibers that transmit light and permit one to see around corners
filly—young female horse prior to her fourth birthday
flag—rhythmic motion of stallion's tail when he ejaculates
float—(v.) to file a horse's teeth to remove sharp points; (n.) a filelike instrument used to float teeth
fly back—a bad habit in which a horse will suddenly pull back, often resulting in a broken halter or tie
foal—(n.) newborn horse of either sex; (v.) the act of foaling, when a mare delivers her young
follicle—fluid-filled blisterlike sack on ovary which contains the ovum (egg)
founder—laminitis, inflammation of the sensory laminae of the hoof
frog—the soft but tough triangular pad of the sole of the horse's hoof
gas colic—colic caused by excessive amounts of gas in the stomach and/or intestines
gaskin—the heavy muscular area between the hock and the stifle
gelding—a castrated male horse
gestation—the period of time between conception and giving birth
girth—belly band—strap around horse's body just behind front legs, which holds saddle or harness in place
glans penis—the end of the penis
granuloma—an excessive amount of non-healing tissue in a wound
gravel—an abscess of the hoof wall extending from the white line to the coronet
"green" or green-broke—a horse not fully trained
grey—a horse with a truly grey coat
grey roan—a horse with a coat of mixed grey and white hairs
gymkhana—western games on horseback
hackamore—a device to guide a horse without a bit, in effect a bitless bridle
halter—harness that fits over the horse's head by which it may be tied or handled
hame—metal or wooden curved piece to fit the collar on a draft harness
hame strap—a short strap which connects the right and left hames together on top and bottom
hand—four inches of height on a horse, measured from its withers to the ground
hand breeding—breeding a mare to a stallion under controlled conditions
hand-twitch—using your hand to hold the horse's nose (as a twitch)
hard keeper—an animal that requires more than the usual amount of food to stay in good condition
haylage—silage made from hay or grass, often referred to as "grass silage"
"heat"—period of estrus when mare will accept stallion
heating—temperature rise as hay or fodder ferments, dries, or cures
hock—tarsal joint between the tibia and cannon, corresponding to human heel
hoof—the hard horny covering of the horse's foot
hoof dressing—a preparation designed to be applied to the hoof either for conditioning or for appearance
hoof dressing—a preparation designed to be applied to the hoof either for conditioning or for appearance.
hoof packing—material, usually claylike, to be applied to the bottom of the horse's hoof
hoof pick—a metal one-tined "rake" to clean debris from a horse's hoof
horse—stallion in race program terminology, an uncastrated male horse past his fourth birthday
horse trailer—trailer used to transport horses

hot horse—a horse sweaty, warm, and puffing from a recent workout
humane twitch—a clamp-type twitch
impaction—blockage of the digestive tract with food material (usually in the large intestine)
interdental space—the space on the horse's jawbone between the incisors and pre-molars where there are no teeth, making room for the bit
intestinal flora—the normal bacteria found in the intestine
intussusception—telescoping of the intestine
isoerythrolysis—a condition in which antibodies in the mare's colostrum destroy the foal's red blood cells
jog cart—two-wheeled cart used to exercise Standardbred horses, heavier than a race sulky
"killer"—the person who butchers horses for meat
knee—the carpal joint, between the radius and the cannon of the foreleg
knee boot—leather or plastic device used to protect knees from bruising each other as horse jogs or races
laminitis—founder, inflammation of the sensitive laminae, or plates of vascular tissue, of the wall of the horse's hoof
larvae—the immature stage of development of the internal parasite following the egg stage
lead rope—a rope usually having a snap on one end, used to lead or tie a horse
lead shank—a webbing or leather strap with short length of chain and a snap, used to lead a horse
legume—a class of plants that manufacture their own nitrogen while growing; alfalfa and clover are the most common
leptospirosis—an infectious disease caused by various leptospira bacteria affecting most warm-blooded species
let down—stopping training, usually done gradually; when milk begins to flow from the mare
line—the strap leading from the bit to the driver's hands in a driving harness (see *reins)*
liniment—a liquid applied externally to increase circulation to a part of the body
lymph—a usually clear fluid similar to blood serum; it may be free in the tissues of the body, in lymph vessels, or part of the blood
lymph node—gland in the body that filters the lymph
manners—the degree of training of a horse in his interactions with humans and other horses
mare—female horse after her fourth birthday
martingale—strap from cavesson to girth to keep horse from throwing its head up, or from hames to girth to help back load
moon blindness—periodic opthalmia, or uveitis (inflammation of internal eye that comes and goes)
navicular—(n.) tiny bone, part of coffin joint; disease of this bone causing lameness
near side—left side
neck yoke—wooden device that holds end of pole up and is attached to hames with chain or strap on a draft harness
New Zealand rug—a particularly heavy, warm horse blanket covered with a layer of canvas
nose clamp—humane twitch
off side—right side
open—not pregnant; a term for the outmoded procedure of reaching into a mare's vagina prior to breeding to open her cervix
open joint—joint opened by a penetrating wound
ovary—one of the pair of ovum (egg)-producing glands of the female, which also produce sex hormones
ovulation—when the follicle ruptures and the ovum, or egg, is released into the oviduct
ovum—egg
oxytocin—the portion of the posterior pituitary hormone which causes milk letdown and contraction of the uterus during foaling
palomino—horse with golden or tan coat and white or cream mane and tail
parade horse—a horse trained to carry ornamented tack in parades
parascarid—the ascarid of the horse
parasite—internal: a living multicelled organism inside another animal, usually intestinal worms; external: an organism that lives on the outside, most usually the louse
parrot mouth—upper incisor teeth extending beyond lower incisors
pastern—area and joint between fetlock and hoof
pasture breeding—when a stallion is pastured with mares and breeding takes place as in the wild
Pelham bit—a combination of snaffle bit and curb bit requiring two reins, used in English riding
performance horse—a horse especially accomplished in showing, jumping, and dressage
periodic opthalmia—see **moon blindness**
placenta—(afterbirth), the membrane attached to the inside of the uterus which takes nutrients from the mare's blood to the fetus through the umbilical cord
points—mane, tail, lower legs, and nose
pole barn—a barn built on poles set in the earth
pony—technically, a horse under 14.2 hands, but for practical purposes, individuals of one of the classic pony breeds such as Shetland, Welsh, Connemara, Pony of America, etc.
Pony Club—a national organization that teaches youngsters to care for and ride horses
P.O.P.—purified oxytocin principal
posterior pituitary extract—hormone produced by the pituitary gland causing milk letdown and contraction of the uterus at foaling
Potomac fever—disease caused by a rickettsia (Ehrlichia equi), with acute projectile diarrhea, laminitis, and usually death; its means of spread from animal to animal has not been determined
pre-potent—a stallion that passes on more than the usual number of traits
progesterone—the hormone produced by the corpus luteum, which helps to maintain pregnancy and control the estrus cycle
proud flesh—protrusion of tissue from wound that will not heal
purebred—an animal with both sire and dam of the same breed (see Throughbred)
put down—to euthanize
quarter—usually refers to the portion of the wall of the hoof such as inside rear quarter, inside front quarter, outside rear quarter, etc.
quarter crack—a split starting at the lower edge of the hoof and running up to the coronet
quarter horse—American Quarter Horse, a muscular type of horse developed for great endurance under saddle and ability to sprint 1/4 mile (hasn't been done with other breeds)
quidding—spitting out pieces of partially chewed hay
quittor—infection of the lateral cartilage of the hoof
rabies—usually fatal virus disease of warm-blooded animals causing paralysis, convulsions, and inability to swallow; usually spread by bites from infected animals
race sulky—light two-wheeled vehicle used in Standardbred racing
radiograph—X ray
reabsorb—possible absorption of an early embryo back into the mare's system
red roan—horse with coat of mixed red and white hair
rein—strap from bit to rider by which horse is controlled (see **line**)
retained placenta—afterbirth that has not been expelled in first three hours after foaling
rhino—short for rhinopneumonitis
rhinopneumonitis—herpes, a viral disease of horse causing respiratory problems ("snots") in young and abortion in pregnant mares

ring bone—arthritis of coffin joint and/or pastern joint causing excessive bone growth
roan—any horse with mixture of white and colored hair
rolling—horse lying down and rolling over, may be normal or result of pain
Rompun—brand of xylazine, an analgesic sedative mixture used as a painkiller, pre-anesthetic, etc.
rug—see **New Zealand rug**
ruminant—animal with four-chambered stomach (cow, sheep, goat, deer)
septicemia—acute generalized infection from virus or bacteria ("blood poisoning")
shafts—a pair of poles that fit on either side of a horse in a single harness
shank—lead shank
sharp teeth—molars that have sharp points that injure tongue or cheek
shedding blade—metal blade with short teeth to scrape out loose hair
sheet—cover for horse made of light canvas or cotton
sheet cotton—cotton pressed into thin sheets, used under leg wraps
shoer—horse shoer, farrier
side step—the maneuver in which a horse moves sideways a step at a time
silage—fodder of higher moisture content than hay stored in airtight structure
single tree—single whiffletree
sire—male parent
sleeping sickness—encephalomyelitis
snaffle bit—simple bit with one joint in center
snip—small white streak above or on nose
"snots"—rhinopneumenitis in foals when thick mucous runs from nose
sock—white above fetlock
sole—the bottom layer of the hoof
spasmodic colic—acute intermittent colic as digestive tract spasms (usually involves small intestine)
stake out—tie an animal on a long rope or chain to a stake driven in the ground
stall chewer—see **wind sucker**
stallion—male horse four years of age and over
stallion syndicate—a financial investment group owning shares in a stallion
Standardbred—an American breed of horse developed for harness racing
standing—when a stallion is at a breeding farm to breed mares brought to him
standing bandage—a bandage held up by wraps down to the hoof
stocking—white leg marking above the cannon
stomach tube—tube passed usually through the nose into the stomach
straight bit—a simple bar bit with no breaks, joints, or projections
straight stall—stall with two walls and manger where horse is tied
strain—the action of a mare when trying to expel the foal, holding her breath and contracting her abdominal muscles
strangles—bacterial respiratory disease caused by *Streptococcus equii* causing swollen abscessed glands
strike—when horse reaches up, out, and down with front foot
stripe—white streak down face
strongyle—"blood worm"
stud—stallion
stud, stud farm—farm where mares are bred
stud fee—the charge for breeding to a stallion
"suck wind"—the action of a mare taking air into her genital tract
sulky—two-wheeled cart
sutured—Caslick operation having been performed
swamp fever—equine infectious anemia
sweat—a mild liniment put on under a waterproof wrapping to "draw" swelling and infection
sweat scraper—metal blade to scrape sweat and water off horse's coat
sweet feed—feed containing molasses
tack—harness, saddles, and equipment
tail rope—rope attached to horse's tail by a half hitch, then tied forward to neck or harness
tail wrap—material to wrap tail during breeding or examination and foaling
tease—the action of a mare in heat; the action of a stallion when he sees mares; to bring a teaser stallion near mares to determine if they are in heat
teaser—a stallion used to determine if mares are in heat
teasing—in heat
test jump—to allow a teaser to mount a mare before the actual breeding stallion is risked
tetanus—bacterial disease caused by *Clostridium tetani*
Texas gate—gate of barbed wire
third eyelid—nictitating membrane, pink membrane in inner corner of eye that can extend across eyeball
Thoroughbred—breed of horse in which all individuals can trace ancestry back to one of five Arabian stallions
thrombus—clot in, or blocking, a blood vessel
thrush—equine pododermatitis, a foul-smelling disease of the frog and surrounding area of the horse's foot caused by *Fusi-bacterium necrophorum*. It is often associated with filth, but a more likely predisposing cause is lack of exercise and neglected foot care
timothy—grass hay, *Phleum pratense*
torsion—intestine twisted off; torsion of uterus in mare in which uterus is twisted shut at neck
trace—heavy strap that attaches the harness to the vehicle or load being pulled
trailer—vehicle towed to move horses; long extension on heel of horseshoe
trainer—person who trains horses usually owned by others
tube—stomach tube
twitch—tool used to restrain horse by pinching nose
udder—mammary gland
uterus—the organ in which the embryo and fetus develop, also referred to as womb or wethers
uveitis—inflammation of internal eye
vagina—sleevelike connection between vulva and cervix
vaginal speculum—instrument to enable dilation of vagina so it and cervix may be examined visually
van—horse truck with large box holding several horses
VEE—Venezuelan equine encephalomyelitis
volvulus—twist of intestinal tract
vulva—external female genitalia
war bridle—restraining rope placed under nose, over gums, and up over poll
wax—colostrum coming from or coagulated on mare's teats just prior to foaling
waxing—showing evidence of wax
weanling—foal between weaning and one year
WEE—western equine encephalomyelitis
wethers—uterus
whiffletree (*whippletree*)—a device, usually wooden with metal rings or hooks, to which traces are attached; may also be double to hold two single whiffletrees
whip training—training horse to respond to touch of whip
white line—white border between sole and wall of horse's hoof (see hoof diagram page 41)
wind sucker—cribber, a horse that holds an object with its teeth and sucks in air
wink—opening and closing of the mare's vulva exposing the clitoris
withers—top of shoulders of horse
wolf teeth—small vestigial first pre-molar
worms—internal parasites
yellow body—*corpus luteum*

SAMPLE HEALTH RECORD

INDIVIDUAL MARE LIFE-TIME HEALTH RECORD

Owner: Sally Jones
Address: Smith Town
Telephone No.: 578 - 3277

ame: Dolly
Sire: Unknown
Dam: Polly
Born: 1970
Tattoo:
Breed: Morgan
Reg. No.: Grade
Color: Bay, Star, snip
Markings: R.F. coronet

Yr.	Pre-Breeding Exam	First Heat	Second Heat	Third Heat	First Service	Sire	Second Service	Sire	Third Service	Sire	Preg.	Due	Date Foaled
1981	4/4/81	3/12?	4/4/	4/25	4/26	Ethan	4/28	Eth..			5/25	3/28/82	4/4/82 - filly
1982	4/14/82	4/13	5/4	5/25	5/25	Ethan					6/24	5/24/83	4/25/83 - Colt

Date

Genital Tract Exam and Treatment

Mo.	Day	Year	Vulva Vagina	Cervix	Uterus	L. O.	R. O.	Treatment	Remarks
4	4	81	n	dom + open	flaccid, n	30B 1¼X1¼	bare 1¼X1¼	Culture	neg. Breed next heat
4	25	81	n	" "	" n	35B	bare		Try Tom.
4	28	81	n	wide open	"	40D	bare	bred again as	soon as practl.
4	30	81	n	closing	firming	pit CH?	bare	recheck	
5	13	81	ne	long firm	firm	CL?	bare	re check @30 day	pregnant
5	24	81			P.U.H.	ne	ne		
6	25	81			good preg.	ne	ne	start rhino series	rex in fall
10	15	81			" "			repeat rhino	
4	14	82	clean no wind	closing	n for 10 days	bare	pit?	was in heat yesterday?	
5	4	82	n	open	n	bare	35B	Culture -	neg
5	25	82	n	open	n	bare	40D	Breed today	
5	27	82	n	ne	n	bare	pit	recheck 18-19 days	
6	13	82	ne	long firm	firm	ne	ne	recheck 6/24	preg
6	24	82			P.R.H.			recheck 30 days	
7	27	82			good preg				
10	13	82			" "			start Rhino	

INDEX

W

Y